A Color Atlas of Comparative Pathology of Pulmonary Tuberculosis

A Color Atlas of Comparative Pathology of Pulmonary Tuberculosis

Edited by
Franz Joel Leong
Véronique Dartois
Thomas Dick

CRC Press
Taylor & Francis Group
Boca Raton London New York

CRC Press is an imprint of the
Taylor & Francis Group, an **informa** business

CRC Press
Taylor & Francis Group
6000 Broken Sound Parkway NW, Suite 300
Boca Raton, FL 33487-2742

First issued in hardback 2017

ISBN 13: 978-1-138-45199-5 (hbk)
ISBN 13: 978-1-4398-3527-2 (pbk)

Library of Congress Cataloging-in-Publication Data

A color atlas of comparative pathology of pulmonary tuberculosis / editors, Franz Joel Leong, Veronique Dartois, and Thomas Dick.
p. ; cm.
Includes bibliographical references and index.
Summary: "Written by notable authorities from the several disciplines involved with tuberculosis research, this book is the first pictorial histopathology atlas with annotations on tuberculosis. The book's drug discovery and animal model perspective makes in applicative rather than academic and presents the material in a highly readable format. Each section will cover the manifestations of induced or acquired pulmonary tuberculosis in one of six animal species (models) - human, non-human primate, rabbit, guinea pig, rat, and mouse. It provides low, medium, and high power microscopy views of lung tissue in color to enhance the understanding of TB for newcomers and the senior tuberculosis researchers."--Provided by publisher.
ISBN 978-1-4398-3527-2 (pbk. : alk. paper)
1. Tuberculosis--Atlases. I. Leong, F. Joel W.-M. II. Dartois, Veronique. III. Dick, Thomas.
[DNLM: 1. Tuberculosis, Pulmonary--pathology--Atlases. 2. Lung--pathology--Atlases. 3. Mycobacterium tuberculosis--Atlases. 4. Pathology, Veterinary--Atlases. WF 17 C7195 2011]

RC311.19.C65 2011
616.9'9500222--dc22 2010016285

Visit the Taylor & Francis Web site at
http://www.taylorandfrancis.com

and the CRC Press Web site at
http://www.crcpress.com

Table of Contents

Preface

THIS WORK BEGAN AS a small collaborative project in 2007, a complement to our drug discovery and clinical development activities. Our institute, the Novartis Institute for Tropical Diseases (NITD), is a center for drug discovery in the areas of tuberculosis (TB), dengue, and malaria. We aim to discover novel treatments and prevention methods for major infectious neglected diseases. In developing countries where these diseases are endemic, Novartis will make treatments available to poor patients without profit.

NITD is also a center for teaching and training of postdoctoral fellows and graduate students, especially young scientists from countries where these diseases are endemic. The institute was founded with funding from Novartis and the Singapore Economic Development Board. Far from being an ivory tower, NITD is able to operate because of the collaborations we have made with academic, clinical, nongovernment, and commercial institutions at all levels. This publication is the product of one such collaboration and without making any grand claims is intended as a simple, useful resource—a visual reference which will allow an appreciation of the histopathological differences of TB between different animal models. It is not intended for consultant histopathologists, but for all scientists and students working in the field of TB.

This atlas provides a visual comparison of histopathological manifestations of TB disease in different animal species and man. However, one has to keep in mind that disease expression in animal models is dependent upon the TB strain used, the number of bacilli for infection, the route of infection, the timing, and the animal strain. Standardization is not possible, and many clinical terms do not have parallels in animal models.

Nonetheless, we hope the images provided are helpful to those involved in research practice.

Joel Leong, M.B. B.S., D.Phil. (Oxon)
Senior Clinical Research Physician

Véronique Dartois, Ph.D.
Head of Pharmacology

Thomas Dick, Ph.D.
Head of TB Unit
Novartis Institute for Tropical Diseases, Biopolis, Singapore
http://www.novartis.com/research/nitd/

Acknowledgments

We would like to acknowledge the support of Novartis, the Singapore Economic Development Board, the Bill and Melinda Gates Foundation, the Wellcome Trust, and the UBS Optimus Foundation.

Additionally, we would like to thank our colleagues, both local and international, for creating the environment in which such an atlas has utility. Without the contributions of our collaborators, the materials necessary for creating the images you see here would not be available.

Finally, we thank family and friends—those who are there for us, and those who remind us that work is not the only thing which defines our humanity and that we will be remembered not just for our achievements, but for how we treated others.

Contributors

El Moukhtar Aliouat, PhD
Department of Parasitology, Faculty of Biological and Pharmaceutical Sciences
University of Lille Nord de France
Lille, France
elmoukhtar.aliouat-3@univ-lille2.fr

V. Balasubramanian, PhD
AstraZeneca India Pvt. Ltd.
Hebbal, Bangalore, India
bala.subramanian@astrazeneca.com

Clifton E. Barry, 3rd, PhD
Tuberculosis Research Section, LCID, NIAID, NIH
Bethesda, MD
cbarry@niaid.nih.gov

Randall J. Basaraba, DVM PhD
Department of Microbiology, Immunology and Pathology
Colorado State University
Fort Collins, CO
basaraba@colostate.edu

Sowmya Bharath, MVSc
AstraZeneca India Pvt. Ltd.
Hebbal, Bangalore, India
sowmya.bharath@astrazeneca.com

Pablo Bifani, PhD
Novartis Institute for Tropical Diseases
Singapore
pablo.bifani@novartis.com

Colette Creusy, MD PhD
Groupe Hospitalier de l'Institut Catholique Lillois (GHICL), Hospital Saint Vincent
University Catholique de Lille
Lille, France
ccreusy@nordnet.fr

Véronique Dartois, PhD
Novartis Institute for Tropical Diseases
Singapore
veronique.dartois@novartis.com

Thomas Dick, PhD
Novartis Institute for Tropical Diseases
Singapore
thomas.dick@novartis.com

Seokyong Eum, PhD
Division of Immunopathology and Cellular Immunology
International Tuberculosis Research Center
Masan, Republic of Korea
syeumkr@itrc.re.kr

JoAnne L. Flynn, PhD
Department of Molecular Genetics and Biochemistry
University of Pittsburgh School of Medicine
Pittsburgh, PA
joanne@pitt.edu

Paul L. Herrling, PhD
Novartis International AG
Basel, Switzerland
paul.herrling@novartis.com

Gilla Kaplan, PhD
Laboratory of Mycobacterial Immunity and Pathogenesis
PHRI Center/UMDNJ
Newark, NJ
kaplangi@umdnj.edu

Klaus Kayser, MD PhD
Institute of Pathology
Charite, Berlin
klaus.kayser@charite.de

Edwin Klein, VMD
University of Pittsburgh School of Medicine
Division of Laboratory Animal Resources
Pittsburgh, PA
eklein@pitt.edu

Anthony S.-Y. Leong, MB.BS, MD, FRCPA, FRCPath, FASCP, FCAP, FHKAM (Pathol), Honorary FHKCPath., Honorary FRCPT
Discipline of Anatomical Pathology
University of Newcastle
Newcastle, Australia
anthony.leong@newcastle.edu.au

F. Joel W.-M. Leong, MB.BS, D.Phil (Oxon)
Novartis Institute for Tropical Diseases
Singapore
joel.leong@novartis.com

Ian M. Orme, PhD
Department of Microbiology, Immunology and Pathology
Colorado State University
Fort Collins, CO
Ian.Orme@colostate.edu

Amit Singhal, PhD
Novartis Institute for Tropical Diseases
Singapore
amit.singhal@novartis.com

Liana Tsenova, MD
Laboratory Mycobacterial Immunity and Pathogenesis
PHRI Center/UMDNJ
Newark, NJ
tsenovli@umdnj.edu

Laura E. Via, PhD
Tuberculosis Research Section, LCID, NIAID, NIH
Bethesda, MD
lvia@niaid.nih.gov

Introduction

Human infection by *Mycobacterium tuberculosis* (MTB) is regarded as one of the so-called "specific" infections. This type of infection induces quite characteristic, though non-disease-specific morphological changes of the infected organ which are called epithelioid granulomas (Cree, 1997). The bacteria are found worldwide and enter the human body via the conducting airways in 80% to 90% of infections. Other primarily infected organs include the intestines and the skin. Tuberculosis can be demonstrated in a broad variety of animals including mammals (*Mycobacterium bovis*), birds (*Mycobacterium avium*), or even fish, where it is a main infectious agent in fish breeding (*Mycobacterium pseudoshottsii*) (Jacobs et al., 2009).

Historically, human tuberculosis has been seen in prehistoric humans of a Neolithic settlement in the Eastern Mediterranean (7000 BC), in Egyptian mummies, or in skeletons of the Paracas-Caverna culture (circa 750 BC to circa AD 100) (Konomi et al., 2002; Gomez i Prat and de Souza, 2003; Zink et al., 2003, Hershkovitz et al., 2008). Tuberculosis has been called "phthisis" by Hippocrates (460–377 BC) and his Greek colleagues, indicating that it was already a disease of social significance at that time.

The details of the infectious pathways and several "therapeutic regimes" have been known for the last 200 years, starting with the investigations of René Theophile Hyancinthe Laennec (1781–1826) which are described in his books "Traité de l'auscultation médiate," and "Traité des maladies du poumon et du coeur." The detection of the tuberculosis bacterium by Robert Koch followed in 1882. The current widely used therapeutic agents, namely ethambutol, isoniazid, and rifampicin, were developed in the middle of the last century shortly after 1950. Newly developed drugs include oxazolidinones (linezolid, PNU-100480), nitroimidazoles (nitroimidazopyran PA-824, metronidazole), 2-pyridone, riminophenazines, and diarylquinolines (Tomioka, 2006). Analysis of the mycobacterial virulence genes and details of the cellular host defense mechanisms, including the activation of killer T cells, have been investigated since the end of the last century (Tomioka, 2006; Hohn et al., 2007).

Thus, man has been fighting against tuberculosis for a long time and has increased his knowledge about the disease and its causes to a high level. Why is tuberculosis still considered to be of significant harm to man? Why is it that about 30% of the world's population is thought to be infected by the tuberculosis bacterium, and why is it that about 1.8 million humans have to die from tuberculosis each year?

The reasons are twofold: In addition to the virulence of the infectious agent, the density of human population in relation to the infectious risk and the status of the host defense system including the therapeutic strategies play a main role in how and to what extent an infection spreads within a population. The development of tuberculosis within the last 10 years is characterized by an increasing number of persons affected by "normal tuberculosis" in developing countries, and a contemporary decrease of these infections in developed countries, which is however, "balanced" by a steady increase of multiresistant and extraresistant tuberculosis, especially in the Western countries. Without any doubt the population of developing countries is "exploding" with unavoidable consequences such as malnutrition or starvation, collapse of hygiene and logistics, as well as clustering of potential tuberculosis victims. For example, malnutrition, exposure to potential harmful airborne substances (fine particulate, smoking, etc.), alcohol consumption, and insufficient housing are suggestive causes for the high rate of tuberculosis in soldiers of the former Soviet republics. International care programs are of limited use only because many of these patients do not take, but sell, the antituberculosis drugs distributed for free.

Increasing globalization, on the other hand, induces a nearly unlimited transfer of infectious agents into and out from the developed countries enhancing the development of multidrug resistance in both the developing and developed countries. Infection with MTB might take months until clinical significance, cannot be noted immediately, and is often hard to diagnose at the beginning, thus allowing multiple infectious contacts with potential victims. The diagnosis itself depends upon the clinical symptoms (which might be weak) and the occurrence of tuberculosis in the individual environment. Usually, a correct diagnosis requires tissue examination, often in combination with expensive molecular genetic analysis (polymerase chain reaction, PCR, with appropriate primers). Whereas the knowledge of characteristic tuberculosis-associated tissue lesions (epithelioid granulomas) is being replaced by more sensitive PCR examinations in developed countries, knowledge of MTB fundamentals is essential for correct diagnosis in an environment which cannot provide expensive diagnosis procedures and therapeutic regimes.

This atlas exemplarily explains the consistency and variety of tuberculosis lesions in human and in animal models. It can be used as a solid basis in tissue examinations in a search for lesions that are characteristic for tuberculosis. In addition, the general spread of pulmonary tuberculosis is shown in detail, allowing a close insight into the pathways of a disease with great social impact. It will, hopefully, serve to allow a firm and consistent diagnosis, which is the prerequisite for a successful and economic treatment of tuberculosis victims.

Klaus Kayser M.D., Ph.D.
Professor of Pathology
Institute of Pathology
Charite, Berlin

REFERENCES

Cree, I.A., ed. 1997. *Pathology,* London, New York: Chapman & Hall Medical.

Gomez, I., Prat, J., and De Souza, S.M. 2003. Prehistoric tuberculosis in America: adding comments to a literature review. *Mem Inst Oswaldo Cruz,* 98 Suppl. 1, 151–159.

Hershkovitz, I., Donoghue, H.D., Minnikin, D.E., Besra, G.S., Lee, O.Y., Gernaey, A.M., Galili, E., Eshed, V., Greenblatt, C.L., Lemma, E., Bar-Gal, G.K., and Spigelman, M. 2008. Detection and molecular characterization of 9,000-year-old *Mycobacterium tuberculosis* from a Neolithic settlement in the eastern Mediterranean. *PLoS One,* 3, e3426.

Hohn, H., Kortsik, C., Zehbe, I., Hitzler, W.E., Kayser, K., Freitag, K., Neukirch, C., Andersen, P., Doherty, T.M., and Maeurer, M. 2007. MHC class II tetramer guided detection of *Mycobacterium tuberculosis*-specific CD4+ T cells in peripheral blood from patients with pulmonary tuberculosis. *Scand J Immunol,* 65, 467–478.

Jacobs, J.M., Stine, C.B., Baya, A.M., and Kent, M.L. 2009. A review of mycobacteriosis in marine fish. *J Fish Dis,* 32, 119–130.

Konomi, N., Lebwohl, E., Mowbray, K., Tattersall, I., and Zhang, D. 2002. Detection of mycobacterial DNA in Andean mummies. *J Clin Microbiol,* 40, 4738–4740.

Tomioka, H. 2006. Current status of some antituberculosis drugs and the development of new antituberculous agents with special reference to their in vitro and in vivo antimicrobial activities. *Curr Pharm Des,* 12, 4047–4070.

Zink, A.R., Sola, C., Reischl, U., Grabner, W., Rastogi, N., Wolf, H., and Nerlich, A.G. 2003. Characterization of *Mycobacterium tuberculosis* complex DNAs from Egyptian mummies by spoligotyping. *J Clin Microbiol,* 41, 359–367.

Abbreviations

AD	Alveolar duct
Al	Alveolus
AS	Alveolar sac
AT	Adipose tissue
Br	Bronchiole, bronchus
BSL	Biosafety Level
Ca	Calcification
CFU	Colony Forming Unit
E	Eosinophil
F	Fibrosis
G	Granuloma
H	Histiocyte, macrophage
HP	High power. Resolution at which nuclei detail can be discerned. At least 20× objective magnification.
H&E	Hematoxylin and eosin stain
L	Lymphocyte
Lg	Langhans' type multinucleate giant cell
LN	Lymph node
LP	Low power. Resolution at which tissue architecture can be appreciated. 1× to 5× objective magnification.
MP	Medium power. Somewhere between 10× to 20× objective magnification.
MTB	*Mycobacterium tuberculosis*
N	Necrosis
PA	Pulmonary artery
PCR	Polymerase chain reaction
PV	Pulmonary vein
TB	Tuberculosis
TBr	Terminal bronchiole
Vs	Vessel
ZN	Ziehl-Neelsen stain

I

Drug Discovery for Neglected Diseases of the Developing World

Paul L. Herrling

CONTENTS

BACKGROUND

The drug discovery process for neglected diseases of the developing world is identical to the process applied to the discovery of medicines for affluent patients. However, the context is not. While drug discovery for diseases like cancer, Alzheimer's disease, diabetes, or AIDS can build on very large bodies of basic science knowledge, this is not the case for diseases that have been neglected for decades such as dengue, tuberculosis (*Mycobacterium tuberculosis,* MTB), malaria (*Plasmodium falciparum* and *Plasmodium vivax*), leishmaniasis (*Leishmania* sp.), diarrheal disease (several infective agents), or Buruli ulcers (*Mycobacterium ulcerans*), filiariasis (*Wuchereria bancrofti, Brugia malayi,* and others), diseases that have predominantly occurred in poorer regions of the planet. A snapshot of published papers recorded in Google Scholar illustrates this very clearly (Table 1.1). The 10 most neglected diseases are listed by the World Health Organization (World Health Organization, 2010b).

The main reasons why many diseases affecting millions of people are neglected are twofold. First, these diseases are infective diseases, and at one point in the last century the success of the antibiotics for major pathogen-induced diseases led to the erroneous assumption that these diseases had been vanquished and needed no more investment. This argument has been invalidated by evolution resulting in increasing resistance to existing antibiotics (e.g., for tuberculosis; World Health Organization, 2008) or antiparasitics (e.g., for malaria; World Health Organization, 2010a). The second reason relates to the fact that more than 95% of all existing medicines have been discovered and developed by commercial organizations on the one hand, and many of the neglected diseases cited above occur in areas where society is too poor to pay a price for them that would justify the research and development investment for a commercial company. This situation prevailed for about 30 years, but recently a welcome trend has emerged: a number of pharmaceutical companies have begun to allocate research and development resources to neglected diseases in a pro bono fashion (i.e., not expecting commercial returns), and in addition some governments and charitable organizations, most notably the Bill and Melinda Gates Foundation and the Wellcome Trust, have collaborated with academia and pharmaceutical companies to establish what were originally called Public Private Partnerships (PPPs) or now Product Development Partnerships (PDPs). The amplitude of this increasing interest for neglected diseases of the developing world has been documented by Mary Moran and her colleagues in two groundbreaking publications about the emerging pipeline for these diseases and the origin and amount of resources invested (Moran, 2005; Moran et al., 2009).

This atlas is intended to make publicly available the results of recent intensive efforts of a team of scientists at the leading edge of modern

TABLE 1.1 Number of Publications for Diseases of Affluent and Poor Patients*

Diseases Affecting both Poor and Affluent Patients	
Search Term	***Number of Hits***
Cancer	3,176,000
Cancer therapeutic	2,020,000
Diabetes	1,610,000
Diabetes therapeutic	1,080,000
HIV	1,220,000
HIV therapeutic	761,000
Diseases Affecting Predominantly Poor Patients, Neglected Diseases	
Search Term	***Number of Hits***
Tuberculosis	924,000
Tuberculosis therapeutic	533,000
Malaria	647,000
Malaria therapeutic	91,900
Chagas	159,000
Chagas therapeutic	17,300
Leishmania	138,000
Leishmania therapeutic	23,400
Dengue	89,200
Dengue therapeutic	11,900
Diarrheal disease	48,900
Diarrheal disease therapeutic	22,900
Filariasis	19,600
Filariasis therapeutic	8,320
Mycobacterium ulcerans	4,590
Mycobacterium ulcerans therapeutic	1,130

* *Google Scholar Advanced Search without date limit, sampled on 27 December 2009.*

tuberculosis research. It is an excellent start to fill some of the lacking basic and drug discovery science in this long neglected area.

THE DRUG DISCOVERY PROCESS

Drug discovery is a highly complex and multidisciplinary activity building on basic scientific knowledge about disease processes and leading to tools to improve the disease addressed. It is typically performed in the laboratories of pharmaceutical companies, while the basic knowledge generation preceding it occurs predominantly in academic laboratories whose culture is ideal for curiosity-driven science but not for the highly structured and project-driven process of drug discovery. In contrast to a large part of biomedical research that is directed at understanding molecular human disease processes and attempting to correct them in the human organism, research for antituberculosis agents (and other antiinfectives) focuses on inhibiting the growth or killing MTB without negatively affecting the human host. The next chapter will focus on the specifics of tuberculosis drug discovery; here the focus is on general aspects of drug discovery with only a few specific points relating to tuberculosis.

THERAPEUTIC TOOLS

These tools can be small synthetic molecules (with molecular weights of less than 500–700), natural compounds of about the same size, or proteins, therapeutic antibodies, vaccines, cell and gene therapies to name a few of the most important ones. Each has advantages and disadvantages.

Despite their bad reputation in the general public, small chemical molecules are very effective medicines, mainly because nature uses small chemical molecules to modulate biological processes. Examples are neurotransmitters and hormones. Drug discovery mimics these mechanisms with either synthetic or natural molecules in order to beneficially influence molecular disease processes. While synthetic small molecules are relatively cheap and easy to make, the chemical universe available to synthetic chemists is still rather limited, and the synthetic molecules resulting are not always suited for biological activity. Natural compounds, however, are the result of millions of years of chemical experiments during the evolutionary process and are optimized for biological activity. The drawback here is that they are structurally often so complicated that they cannot be easily synthesized. Biological production processes are often the only way to produce them in sufficiently large quantities, making them expensive. Both synthetic and natural small molecules can reach every compartment in the human body, both extra- and intracellular.

Most therapeutic tools contemplated as antituberculosis agents are either synthetic or natural compounds. Proteins, in particular monoclonal antibodies, can be used as therapeutics with very high affinity and specificity, and they are increasingly easy to produce and in general have a more benign side effect profile than small molecules because of their specificity.

There are a rapidly increasing number of therapeutic antibodies in clinical use. Their major drawback is that they are limited to extracellular compartments and must be given parenterally. However, the frequency of their application is as low as once every few months. Vaccines are an excellent and in many cases very efficient way to stimulate human immunological defenses against infective agents; however in tuberculosis this strategy has met with only limited success in the past.

This article will not address other therapeutic tools such as genetic or cell therapy because they are not relevant (yet) for neglected diseases.

DRUG DISCOVERY PHASES

The drug discovery phases (Figure 1.1) are designed to find and develop potential therapeutic tools and to characterize them in sufficient detail to allow their effective and safe clinical use. The process starts with the D0 phase and ends with a successful proof-of-concept or mechanism in Phase IIa (PhIIa). Phase IIb (PhIIb) and Phase III (PhIII) are called full development. The process looks linear, but it is highly iterative and contains a large number of parallel elements. This is indicated in Figure 1.1 by the backward arrows starting at different phases where projects are stopped and the findings fed back to backup projects earlier in the process. In this article only a very short summary can be given. For a detailed discussion, the book by H.P. Rang is recommended (Rang, 2006).

D0

This phase covers all the basic science leading to a better understanding of the disease process and allowing, in the best cases, selection of an effective molecular therapeutic target in a relevant disease process, usually

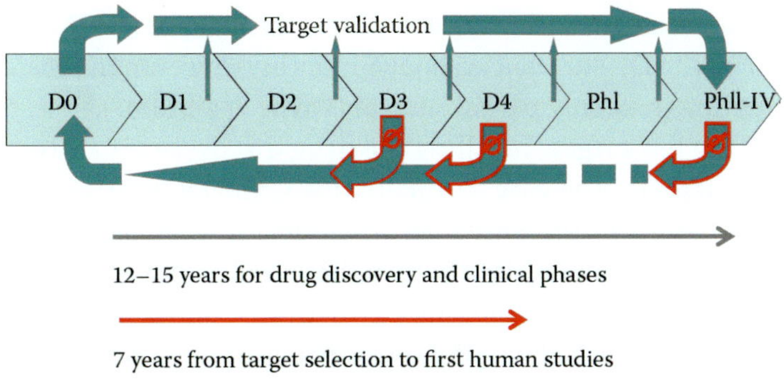

FIGURE 1.1 Phases of the drug discovery process.

an intracellular pathway or a cell-membrane receptor. This phase is not included in the drug discovery process in the strict sense and happens predominantly in academic laboratories, while the following phases occur in the laboratories of pharmaceutical companies. But in the case of neglected diseases there is often an insufficient scientific literature (see above), so that some of the basic science is contributed by the laboratories normally involved in drug discovery. A case in point is the discovery of MTB culture-media effects discovered in our own laboratories (Sequeira et al., 2010) that on the one hand resulted in many time-consuming false positive antibacterial compound classes, but also lead to a better understanding of bacterial physiology that might result in better drug discovery strategies for tuberculosis.

This basic science part of the process has been vastly accelerated in the last 20 years by advances in biomedical sciences and technologies but can still be of very varying length, from a few to dozens of years.

D1

In all cases where the disease process is known in sufficient molecular detail, it is possible to select a "target" or the exact molecule in the disease process whose modulation by a drug should lead to the desired therapeutic effect. At this point the target is only partly validated, for example, by genetic knock-out experiments in which inactivation of the gene coding for the putative target protein leads to the desired therapeutic effect. A target is only fully validated if its modulation results in the desired therapeutic effect in human beings. Target validation during the discovery process is only gradual, and validation data is accrued incrementally (see Figure 1.1). If such a partly validated target is available, the D1 phase is triggered where an assay system is developed allowing the measurement of interactions of candidate molecules with the target in a high-throughput way (Stoeckli and Haag, 2006). The D1 duration is about 6 months. However, in tuberculosis, this strategy has had only partial success (Payne et al., 2007).

If no or too few targets fulfilling the drug discovery criteria are available, as is currently the case in tuberculosis, the alternative strategy is to measure the effect of candidate drugs on the growth of bacteria in varying cell culture conditions which, however, is not straightforward and rather slow because of the small growth rate of MTB. Furthermore, the experiments with this organism need to be carried out in a Biosafety Level 3 (BSL-3) environment (Centers for Disease Control and Prevention, 2000).

D2a

In this phase all available compounds in a company's library are tested in the high-throughput assay or a selection in cellular assays. Compounds interacting with the target or reducing bacterial growth at a predefined potency are called "hits."

D2b

All hits are evaluated by medicinal chemists for "druggable" properties such as solubility, potential toxic moieties, metabolic stability, etc., and the most promising ones are declared "leads." The D2 phase lasts on average about 1.5 years.

D3

In this phase the leads are systematically chemically modified to improve the drug properties, in particular potency, selectivity, and pharmacokinetic parameters such as bioavailability, in addition to the properties described above. The resource requirements in this phase are significantly increased because many chemists are needed to derivatize leads, and pharmacokinetic and dynamic studies in whole animals are time consuming. This phase typically lasts 18 months and also includes upscaling of the chemical production process from a few milligrams to a kilogram scale that will be sufficient for the animal models to follow.

D4

At the completion of D3, efficacy studies in whole animals are initiated, if possible, in two species. In tuberculosis only rodents are practical at this stage, although nonhuman primates show most of the characteristics of the disease (see Part II of this atlas). The first formal toxicological studies in whole animals are performed in this phase and are often the cause of project terminations. Because this is the last phase before human testing, decisions about salt forms and galenic formulation must occur here as well as the elaboration of the strategy to evaluate the scientific proof of concept/mechanism in humans. The D4 phase lasts about 2.5 years.

Phase I–Phase IIa

If the animal studies support it, the new drug candidate is now tested in human volunteers for tolerance, followed by the proof of concept/mechanism. If the mechanism of action of a new drug is sufficiently elucidated, it is often possible to determine if the scientific hypothesis elaborated in in vitro and animal systems is valid in human beings, either healthy volunteers or patients. This proof of concept is nowadays used as a major go/no go decision point and can save significant resources that might be wasted in later expensive clinical phases. Drug discovery activities typically stop at the successful completion of this phase. If unsuccessful, the process is started again incorporating the lessons learned.

Phase IIb–Phase III

In these phases of full development, the exact therapeutic doses are determined and the efficacy and competitive advantage of a new drug documented in double blind studies. In tuberculosis, clinical studies are very protracted, because treatment lasts over 6 months, and relapse probability is evaluated over several years. If these are successful, the new therapeutic is submitted for registration, and the postregistration follow-up is done in Phase IV studies.

In the years since the establishment of the Novartis Institute for Tropical Diseases (NITD), an increasing amount of resources have been allocated toward searching for new antituberculosis agents that should overcome the increasing resistance to old antibiotics observed globally and that hopefully will also achieve clearance of MTB in significantly less time than is currently the case. The following chapters cover some aspects of our experiences in tuberculosis drug research from an animal model and histopathological perspective.

REFERENCES

Centers for Disease Control and Prevention. 2000. Laboratory biosafety level criteria [online]. Centers for Disease Control. Available from: http://www.cdc.gov/OD/ohs/biosfty/bmbl4/bmbl4s3.htm [Accessed January 7, 2010].

Moran, M. 2005. A breakthrough in R&D for neglected diseases: New ways to get the drugs we need. *PLoS Med,* 2, e302.

Moran, M., Guzman, J., Ropars, A.L., McDonald, A., Jameson, N., Omune, B., Ryan, S., and Wu, L. 2009. Neglected disease research and development: How much are we really spending? *PLoS Med,* 6, e30.

Payne, D.J., Gwynn, M.N., Holmes, D.J., and Pompliano, D.L. 2007. Drugs for bad bugs: Confronting the challenges of antibacterial discovery. *Nat Rev Drug Discov,* 6, 29–40.

Rang, H.P., ed. 2006. *Drug discovery and development: Technology in transition.* Philadelphia: Churchill Livingstone Elsevier.

Pethe, K., Sequeira, P.C., Agarwalla, S., Rhee, K., Kuhen, K., Fong, W.Y., Patel, V., Beer, D., Walker, J.R., Duraiswamy, J., Jiricek, J., Keller, T.H., Chatterjee,

A., Tan, M.P., Ujjini, M, Rao, S.P.S., Camacho, L., Bifani, P., Mak, P.A., Ma, I., Barnes, S.W., Chen, Z., Plouffe, D., Thayalan, P., Ng, S.H., Au, M., Lee, B.H., Tan, B.H., Ravindran, S., Nanjundappa, M., Lin, X., Goh, A., Lakshminarayana, S.B., Shoen, C., Cynamon, M., Kreiswirth, B., Dartois, V., Peters, E.C., Glynne, R., Brenner, S., and Dick, T. 2010. A chemical genetic screen in *Mycobacterium tuberculosis* identifies carbon-source dependant growth inhibitors deprived of in vivo efficacy. *Nature Communications* (submitted for review).

Stoeckli, K., and Haag, H. 2006. High-throughput screening. In *Drug discovery and development: Technology in transition,* ed. H.P. Rang, 99–120. Philadelphia: Churchill Livingstone Elsevier.

World Health Organization. 2008. *Anti-tuberculosis drug resistance in the world. Fourth global report. The WHO/IUATLD global project on anti-tuberculosis drug resistance surveillance 2002–2007.* Geneva: World Health Organization.

World Health Organization. 2010a. *Drug resistance: malaria* [online]. World Health Organization. Available from: http://www.who.int/drugresistance/malaria/en/ [Accessed January 9, 2010].

World Health Organization. 2010b. *Neglected tropical diseases* [online]. World Health Organization. Available from: http://www.who.int/neglected_diseases/diseases/en/ [Accessed January 9, 2010].

CHAPTER 2

Tuberculosis Biology and Drug Discovery

Thomas Dick

CONTENTS

MYCOBACTERIUM TUBERCULOSIS (MTB) is the causative agent of tuberculosis (TB). First discovered in 1882 by Robert Koch, the tubercle bacillus has an unusual waxy coat primarily made up of mycolic acids. Because of its unique cell wall structure, the organism does not retain the usual bacteriological stains and thus is neither a Gram positive nor a Gram negative. Ziehl-Neelsen staining is used instead to detect MTB. Another hallmark of the pathogen is that it grows very slowly compared to other bacteria. Whereas *E. coli* divides once every 20 minutes, MTB divides once a day. The size of the MTB genome, about 4,000 genes, is similar to that of *E. coli*. The tubercle bacillus is an aerobic bacillus (i.e., it requires oxygen for growth and is rod shaped). Simple culture media are available to grow MTB in the lab. So is a whole range of experimental animal models, from mice to monkeys, to grow the bacillus in vivo. MTB is a facultative intracellular parasite. It can grow extracellularly as well as intracellularly, both in culture and inside human lesions. Due to the serious disease it can cause, and its way of transmission via aerosol formation, a Biosafety Level 3 facility is required to work with the bacillus (Cole et al., 1998).

DISEASE MANIFESTATIONS AND TREATMENT

MTB is an obligate human parasite that attacks largely the lungs, although other organ systems can be affected such as the central nervous or the lymphatic system and bones and joints. Classical symptoms of pulmonary TB are chronic cough, fever, night sweats, and weight loss. Infection occurs via inhalation of aerosol droplets. Upon entry into the lungs, the pathogen is taken up by alveolar macrophages where it can multiply, disseminate, and establish foci of infection and lesions in other parts of the lungs. Lesion formation usually contains but does not eradicate the bacilli, resulting in asymptomatic latent infection. One in ten latently infected people develop active disease during their lifetime, which includes increased multiplication, spread of the bacterium, and often the formation of liquefied cavities which are open to airways, with the symptoms mentioned above (Stewart et al., 2003).

An effective vaccine is not available for TB. Chemotherapy has been available for about half a century. The current treatment of active TB disease consists of a combination therapy of four drugs, rifampicin, isoniazid, pyrazinamide, and ethambutol, and requires 6 to 9 months to achieve cure of drug-susceptible TB. Multidrug-resistant TB (MDR-TB), defined as resistance against the two most potent drugs, rifampicin and isoniazid, requires the application of second-line drugs, which are in general less effective and more toxic. MDR-TB treatment takes 18 to 24 months to achieve cure. In the case of extensively drug-resistant TB (XDR-TB), defined as MDR-TB plus resistance against our best second-line drugs, fluoroquinolones and an injectable anti-TB drug (i.e., aminoglycoside or capreomycin), treatment is even more difficult and often results in failure and death. Prophylactic treatment of latent TB is currently done with isoniazid for 9 months, but results in only partial eradication. The extremely long and complex treatment regimens for the different forms of TB represent a key problem in global TB control. As most TB cases occur in resource-limited countries, implementation and compliance is a major issue, fueling selection and development of more drug-resistant TB—and human suffering and death (Donald and van Helden, 2009).

A LARGELY UNMET MEDICAL NEED

TB represents a tremendous global problem. Estimates suggest that about 2 million people die of TB every year. Each year sees an additional nine million new TB cases, with about 500,000 MDR-TB patients. The number of XDR-TB infections is increasing. About a third of the human population is thought to be latently infected. In other words, 2 billion people carry MTB without symptoms (and without being infectious), however being in danger of developing active TB disease. HIV coinfection, weakening the immune system, increases the chances of progression to active TB dramatically, resulting in catastrophic situations such as seen, for instance, in Africa (Donald and van Helden, 2009).

The increase of drug-resistant TB clearly demands the discovery and development of new antimycobacterials. At the time of writing, a few new drug candidates—two novel nitroimidazoles and an ATP synthase inhibitor—entered clinical development. While this is very good news, the overall pipeline for new anti-TB drugs remains thin, and clearly more sustained efforts are needed to identify attractive lead compounds and carry out lead optimization to deliver candidates for development (Sacchettini et al., 2008; Nathan, 2009).

Developing new antimycobacterials to keep drug-resistant TB at bay is the primary objective for TB drug discovery. However, there is a second objective that is critical if we are ever to hope to reduce the global TB burden in a significant way: to come up with *better* drugs than the ones we have! Current TB chemotherapies take many months or even years to achieve cure. In other words, our current anti-TB drugs are not *really* effective. This becomes obvious if one considers that common upper respiratory tract infections, caused for instance by *Streptococcus,* can be cured within a week or two. Why does it take days to kill *Streptococcus* but months or years to kill MTB, both residing in the lungs? The potencies of antimycobacterials against tubercle bacilli in the test tube are comparable to the potencies of antibiotics directed against other bacteria. Why do drugs that show good potency against cultured TB bacilli take such a long time to kill the pathogen in patients? The short answer is: we do not know. But some working models have emerged over the past decade. They are centered around the fact that TB is a lesion-based disease and that lesions in TB patients come in vastly different forms (size, structure, contents, immune status, bacilli load), even within the same patient.

THE CHALLENGE OF TB DRUG DISCOVERY

Microbiology textbooks teach us that bacterial growth depends on the culture conditions. For instance, if one takes away oxygen from an obligate aerobe bacterium, it cannot grow. If there is nutrient limitation or if the pH is "wrong," bacteria stop growing. The nature of the largely varying lesion types that MTB generates as its "culture vessels" in humans suggests that the bacillus encounters very different microenvironments in patients. Some are supportive of growth, others not. And indeed, recent work on TB animal models showed that some lesions in some animal models are, for instance, hypoxic, suggesting the bacilli in these lesions cannot grow. Importantly, when nongrowing bacteria in culture are exposed to antibiotics that kill the growing form of the organism, they show "phenotypic" drug resistance (as opposed to genetic drug resistance, due to mutations). Antibiotics usually lose their cidal activity against nonreplicating organisms. That is not at all surprising, because antibacterials were identified and selected based on their activity against the growing form and are thus largely directed against bacterial functions essential for growth, such as macromolecular synthesis machines. Taken together, one working hypothesis is that certain TB lesions present microenvironments that are not supportive of growth, and that the bacilli in these microenvironments stop growing and become "phenotypically" drug resistant. Lesions are known to be dynamic structures. The microenvironments can change over time, and the bacilli that survived the onslaught of drugs in a nongrowing state can resume growth, hence explaining the long treatment time required to cure TB (Barry et al., 2009; Dartois et al., 2009).

That MTB resides in lesions deep inside human tissue might cause another problem, this time for the drugs. Can the drugs actually reach the bacilli inside the lesions at the concentrations required to exert their antimicrobial activity? Very little is known about the penetration of anti-TB drugs into different lesion types in people and even less is know regarding lesion penetration in TB animal models. Some evidence suggests that penetration into lesions is drug and lesion type dependent. Thus we have two working models explaining why our current drugs might not work effectively in patients. Both have to do with the fact that MTB resides in lesions. Some lesions might simply not allow effective penetration of drugs. Those subpopulations of bacilli (whether growing or not) that do not see the drugs are safe simply due to their location. In other lesions (or the same), MTB might face microenvironmental conditions that do not support growth. Even if the drugs do reach the bacilli, they do not affect the pathogen because it is in a nonsusceptible physiological state. These two "special MTB subpopulation" models, explaining the loss of potency of anti-TB drugs when moving from the test tube to patients, are obviously not mutually exclusive (Barry et al., 2009; Dartois et al., 2009).

AN ATLAS FOR TB DRUG DISCOVERY

Why is this atlas on comparative TB histopathology important for TB drug discovery? We do not know which lesion types are the most "difficult" to treat in human TB. What *are* the lesion types that are difficult to penetrate, which lesions contain phenotypically drug-resistant bacilli? What are the growth-terminating "culture conditions" in those lesions? The definition of lesion-specific microenvironmental conditions is key for the development of predictive in vitro MTB culture assays, critical for the discovery of new drugs. To study and understand TB lesions and identify the "difficult" ones, it is required to work on human and animal lesions in parallel. Working on human lesions alone is not sufficient because of limited sample availability (they come from rare surgeries) and because of experimental limitations. Once we have identified the difficult lesion types in man, we can identify the appropriate animal model(s) that displays these lesion types to determine the efficacy of new compounds against bacilli present in each particular lesion type. We would have a predictive animal model for discovering more effective drugs! For that, of course, we have to know, to understand in detail, the different lesion types present in the various animal models. At the moment only the mouse is used widely in efficacy testing, because it is the easiest, fastest, and most cost-effective model—and proven to work. All existing anti-TB drugs work in that model (they were of course also identified there). However, the TB mouse model does not show much of human-like TB lesions.

Work on comparative lesion studies has just begun. This atlas hopes to contribute to that newly emerging field by discussing comparative TB histopathology. What kinds of lesions are present in which TB animal models and in man? That is a start.

REFERENCES

Barry, C.E., 3rd, Boshoff, H.I., Dartois, V., Dick, T., Ehrt, S., Flynn, J., Schnappinger, D., Wilkinson, R.J., and Young, D. 2009. The spectrum of latent tuberculosis: Rethinking the biology and intervention strategies. *Nat Rev Microbiol*, 7, 845–855.

Cole, S.T., Brosch, R., Parkhill, J., Garnier, T., Churcher, C., Harris, D., Gordon, S.V., Eiglmeier, K., Gas, S., Barry, C.E., 3rd, Tekaia, F., Badcock, K., Basham, D., Brown, D., Chillingworth, T., Connor, R., Davies, R., Devlin, K., Feltwell, T., Gentles, S., Hamlin, N., Holroyd, S., Hornsby, T., Jagels, K., Krogh, A., Mclean, J., Moule, S., Murphy, L., Oliver, K., Osborne, J., Quail, M.A., Rajandream, M.A., Rogers, J., Rutter, S., Seeger, K., Skelton, J., Squares, R., Squares, S., Sulston, J.E., Taylor, K., Whitehead, S., and Barrell, B.G. 1998. Deciphering the biology of *Mycobacterium tuberculosis* from the complete genome sequence. *Nature,* 393, 537–544.

Dartois, V., Leong, F.J., and Dick, T. 2009. TB drug discovery: Issues, gaps and the way forward. In *Antiparasitic and antibacterial drug discovery: From molecular targets to drug candidates*, ed. P. Selzer, 415–440. Weinheim: Wiley-VCH.

Donald, P.R., and Van Helden, P.D. 2009. The global burden of tuberculosis—Combating drug resistance in difficult times. *N Engl J Med,* 360, 2393–2395.

Nathan, C. 2009. Taming tuberculosis: A challenge for science and society. *Cell Host Microbe,* 5, 220–224.

Sacchettini, J.C., Rubin, E.J., and Freundlich, J.S. 2008. Drugs versus bugs: In pursuit of the persistent predator *Mycobacterium tuberculosis. Nat Rev Microbiol,* 6, 41–52.

Stewart, G.R., Robertson, B.D., and Young, D.B. 2003. Tuberculosis: A problem with persistence. *Nat Rev Microbiol,* 1, 97–105.

CHAPTER 3

Immunopathology of Tuberculosis Disease across Species

Véronique Dartois

CONTENTS

Mycobacterium tuberculosis (MTB), the etiological agent of human TB, is closely related to all other members of the MTB complex, many of which are host adapted to a number of animal species. Strains of the MTB complex naturally infect a variety of mammals from rodents to cattle (Smith et al., 2009). Hence many animal models of pulmonary MTB infection have been developed over the past decades, by varying the bacterial strain/species, animal species, size, and route of pathogen inoculation. Though most of these species would not naturally develop MTB-induced TB disease, these models have proven useful for the study of TB pathogenesis and immunopathology and for the preclinical testing of drugs and drug combinations. MTB infections of mice, rats, guinea pigs, rabbits, and nonhuman primates are the most commonly used animal models of human tuberculosis.

The pathology of TB disease is driven by the immune response to MTB invasion more than by direct pathogen-inflicted damage. Following transmission through aerosols, a few bacilli are inhaled and penetrate deep into the lung. In the alveolar sacs, MTB is engulfed by alveolar macrophages, key cells of the innate immune system that act as a front-line barrier against infection. The bacilli have a range of systems to circumvent the antibacterial properties of the macrophage, allowing them to reside and replicate in a protected intracellular niche. As the bacterial load increases, T lymphocytes and other immune cells are recruited to the site of infection, where they attract and activate more macrophages in an attempt to eradicate the pathogen. This escalation of the immune response leads to the generation of an organized granulomatous lesion, or tubercle, the hallmark of TB pathology. These structures classically contain concentric layers of immune cells of various types and functions, the center of which can become necrotic. This necrotic core is also called "caseum" due to its cheese-like appearance resulting from the combined lysis of both host cells and pathogens. As they age and develop, granulomas can surround themselves with layers of fibroblasts in an attempt to further contain the pathogen and heal inflamed or damaged lung tissues. One of the most critical steps in disease progression is the appearance of cavities, which result from the encounter of a growing necrotic granuloma with an airway. Rupture of these cavities, which contain high numbers of extracellular tubercle bacilli, enables aerosol transmission, an essential step in the infection cycle of MTB as an obligate pathogen. Most infected individuals can successfully contain the pathogen in early cellular lesions, leading to a state of latent infection with no obvious clinical signs. However, 5% to 10% of latently infected individuals will later develop active disease.

Such a complex interplay between host and pathogen inevitably implies that different animal species will reach different stages of granuloma formation, in terms of structure and composition, leading to varying abilities to contain the pathogen. Currently, not one model fully mimics the complete and elaborate spectrum of lesion types that we see in humans, though it is clearly recognized that the nonhuman primate best reproduces human disease progression and diversity. As our understanding of the TB pathology and latent disease in humans has improved, we have come to realize that latent TB is characterized by a continuum of more or less quiescent and healed lesions rather than a unique and well-defined lesion type (Barry et al., 2009; Young et al., 2009). Current studies of latency in the nonhuman primate aim to determine whether the model can truly and completely reproduce the spectrum of latent disease as defined in the clinic (Lin et al., 2009). Successful use of any model depends on our ability to identify the features and lesion types which they lack or exhibit, recognize their respective limitations, and determine whether these traits are compatible with the expected outcome of the intended study (Basaraba, 2008). The information provided in this atlas was compiled in an attempt to help TB scientists distinguish between the features of all major animal models available and use them with their strengths and limitations in mind.

OVERVIEW OF SPECIES SPECIFIC CHARACTERISTICS

The Mouse

Both logistic and scientific considerations play an important role in the selection of an animal model. For practical reasons, the most popular model so far is the mouse due to its low cost, the availability of genetically defined mouse strains, an extensive literature regarding mouse immunology, and the availability of immune reagents. Most TB drugs and drug combinations which are effective to treat human TB have proven efficacious in the mouse. Provided care is taken to optimize the experimental conditions, the mouse has produced reliable data on the bactericidal (killing rather than inhibiting growth) and sterilizing (eradicating viable bacilli to prevent relapse) activity of existing antituberculosis drugs and informed numerous clinical trials (Orme, 2003; Nuermberger, 2008). However, despite some similarities in the immune control of TB in mice and humans, the progression of disease is markedly different. Outbred wild-type mice are resistant to tuberculosis, while aerosol infection of IFN-γ knockout mice with MTB results in very rapid disease progression, along with granuloma formation throughout the lungs, with little evidence of hematogenous spread to the apical lobes. The granulomas that develop in mice are not the well-formed structures that are observed in humans, but consist of aggregates of lymphocytes and macrophages that do not progress to caseation and liquefaction and therefore lack the hypoxic environment found in the necrotic core of human lesions (Boshoff and Barry, 2005; Via et al., 2008). Finally, cavity formation does not occur in the mouse, while it is generally considered as the lesion type which is most difficult to sterilize. Since the early days of anti-TB therapy, the presence and extent of cavitary disease have often been cited as correlates of poor clinical outcome, development of resistance, and relapse (Aber and Nunn, 1978; Chang et al., 2004). After the onset of acquired immunity, mice establish a chronic phase of the infection, but unlike humans, they maintain high bacillary loads without apparent disease for months. This combination of high bacterial loads with no clinical signs is not seen in an infected human being.

Overall, the mouse model recapitulates a very limited set of the clinicopathological manifestations of tuberculosis in humans. Yet the current strategy for introducing new or existing drugs in TB clinical trials largely relies on extensive murine studies designed to evaluate the bactericidal and sterilizing activity of a drug candidate or drug combination. Although its predictive value is a matter of intense debate, the mouse remains the best-characterized and most economical animal model for experimental chemotherapy.

The Rat

Limited recent findings with the Wistar rat suggest good reproduction of disease progression seen in humans, comparable immune responses and disease control, and partial similarities with the human TB pathology. However, the histological features are not those of typical human tuberculosis. Granulomas have loosely defined borders and lack a trilaminar appearance due to the absence of necrosis and fibrosis (Pablo Bifani, unpublished). The American cotton rat also presents granulomatous disease, often with central necrosis in the lungs, spleen, and lymph nodes of infected animals (Elwood et al., 2007). The cotton rats however, as well as several other rat species investigated to date (Sugawara et al., 2004a,b; Sugawara et al., 2006), do not seem able to control the infection as the Wistar do.

It thus appears that the long-forgotten rat model offers the potential of a reasonably predictive tool in a species equally suited and widely used in the pharmaceutical industry for toxicology and pharmacokinetic studies. Correlations of pharmacodynamic endpoints and drug efficacy with pharmacokinetic profiling and toxicity readouts constitute a marked advantage for preclinical testing of new drug candidates. From a practical standpoint, the rat remains affordable, easy to house under Biosafety Level 3 (BSL-3) conditions, with a growing number of immune reagents and kits becoming commercially available. Further characterization is required to bring this newly revived model to the central TB stage.

The Rabbit and the Guinea Pig

Dannenberg (Dannenberg, 2006) has published and summarized a great deal of detailed characterization of the rabbit model of tuberculosis, though most of the literature describes bovine tuberculosis because susceptible rabbits which were the subject of early classic studies by Lurie have been lost (Allison et al., 1962). Nevertheless, both rabbits and guinea pigs are naturally susceptible to MTB (Lurie, 1949; Turner et al., 2003). Furthermore, clinical MTB strains of higher virulence were recently shown to establish a progressive infection in the rabbit (Sinsimer et al., 2008). In guinea pigs, progressive infection can result from the implantation of a single virulent tubercle bacillus, and all infected guinea pigs will invariably progress to fatal disease, thereby making them more susceptible than humans (Smith and Harding, 1977). Both models more closely mimic the gross pathology of human disease than murine species, in that the mechanisms of granuloma formation with associated caseation are very similar in these hosts and result in the development of regions of low oxygen tension in the necrotic core of caseating lesions (Dannenberg, 2006). In addition, rabbit granulomas can proceed to liquefaction and cavity formation (Dannenberg, 2006). In these animal models, hematogenous spread to uninfected lobes occurs and has been studied intensively. In guinea pigs, the disease is characterized by an exponential increase in bacterial numbers in the lungs before the onset of acquired immunity and extensive tissue destruction during cell-mediated immunity, which ultimately results in the death of the animal. The dermal hypersensitivity reaction to mycobacterial antigens (i.e., the classical tuberculin skin test) in an infected rabbit, and to a lower extent in an infected guinea pig, is an accurate reproduction of dermal responses in infected humans.

Expense of maintenance, cost of purchase, BSL-3 housing, and availability of immunological reagents are more problematic for these two species, even more so for the rabbit. Although reagent development has progressed in the past few years, reagents for flow cytometry studies, cytokine measurement, and immunohistochemistry remain scarce, with limited hope for significant improvement due to the small market.

The Nonhuman Primate

There are many logistical and practical hurdles associated with the use of nonhuman primates: cost of breeding, expense and difficulty in maintaining animals in BSL-3 facilities, and reluctance of bioethical and animal welfare committees to approve monkey studies.

Despite these numerous caveats, the monkey remains a highly relevant species in which to study TB (McMurray, 2000). It provides an excellent representation of human tuberculosis in terms of disease progression, variability of disease outcome, immuno- and histopathology (Lin et al., 2006). Low-dose MTB infection in cynomolgus macaques results in a spectrum of disease similar to that of human infection: primary disease, latent-like infection, and reactivation tuberculosis (Lin et al., 2009). Animals with no obvious disease have no signs of radiographic lung involvement, and many remain free of clinical signs of disease for up to two years. However, lung necropsy of infected but symptom-free animals reveals granulomas, some with caseation. The structure of nonhuman primate granulomas is microscopically and immunologically similar to human granulomas, with hypoxic conditions found across the core. So this nonhuman primate model mimics the development of disease in humans more closely than any other animal model. In addition, many human immunological reagents cross-react with their nonhuman primate counterpart, thus facilitating research and disease characterization in this model.

A MACROSCOPIC VIEW OF TB-INFECTED LUNGS ACROSS SPECIES

Figure 3.1 illustrates the dramatic differences in severity and diversity of pathology across species. Only the nonhuman primate (Figure 3.1e) appears to reproduce the variety and size of lesions seen in the human lung (Figure 3.1f). Note that the dimensions of NHP and human lesions can reach the size of an entire mouse lung. The rat, guinea pig, and rabbit (Figure 3.1b,c,d) show well-defined nodules at the surface of the lung, with most rabbit lesions displaying a caseous necrotic center not visible in the rat and guinea pig. The mouse fails to show well-defined visible granulomas (Figure 3.1a). The drastic differences which can be visualized in this simple macroscopic view of infected lung tissues from the various species reflect the detailed histopathological findings presented in subsequent chapters.

FIGURE 3.1 (opposite page) Macroscopic appearance of TB-infected lungs from mouse (a), rat (b), guinea pig (c), rabbit (d), nonhuman primate [NHP] (e), and human (f) origin. Mouse, rat, and guinea pig were infected with MTB strain H37Rv, the rabbit was infected with Beijing strain HN878, and the monkey with MTB Erdmann. Panel (f) shows lung tissue resected from a patient with difficult-to-treat multidrug-resistant tuberculosis of the Beijing family. The time postinfection varies between 30 days and 6 months and is unknown for the human case. (Mouse, rat, guinea pig, rabbit, NHP, and human photos were provided courtesy of Laura Via, Maxime Hervé, Angelo Izzo, Gilla Kaplan, Edwin Klein, and Seok Yong Eum, respectively.)

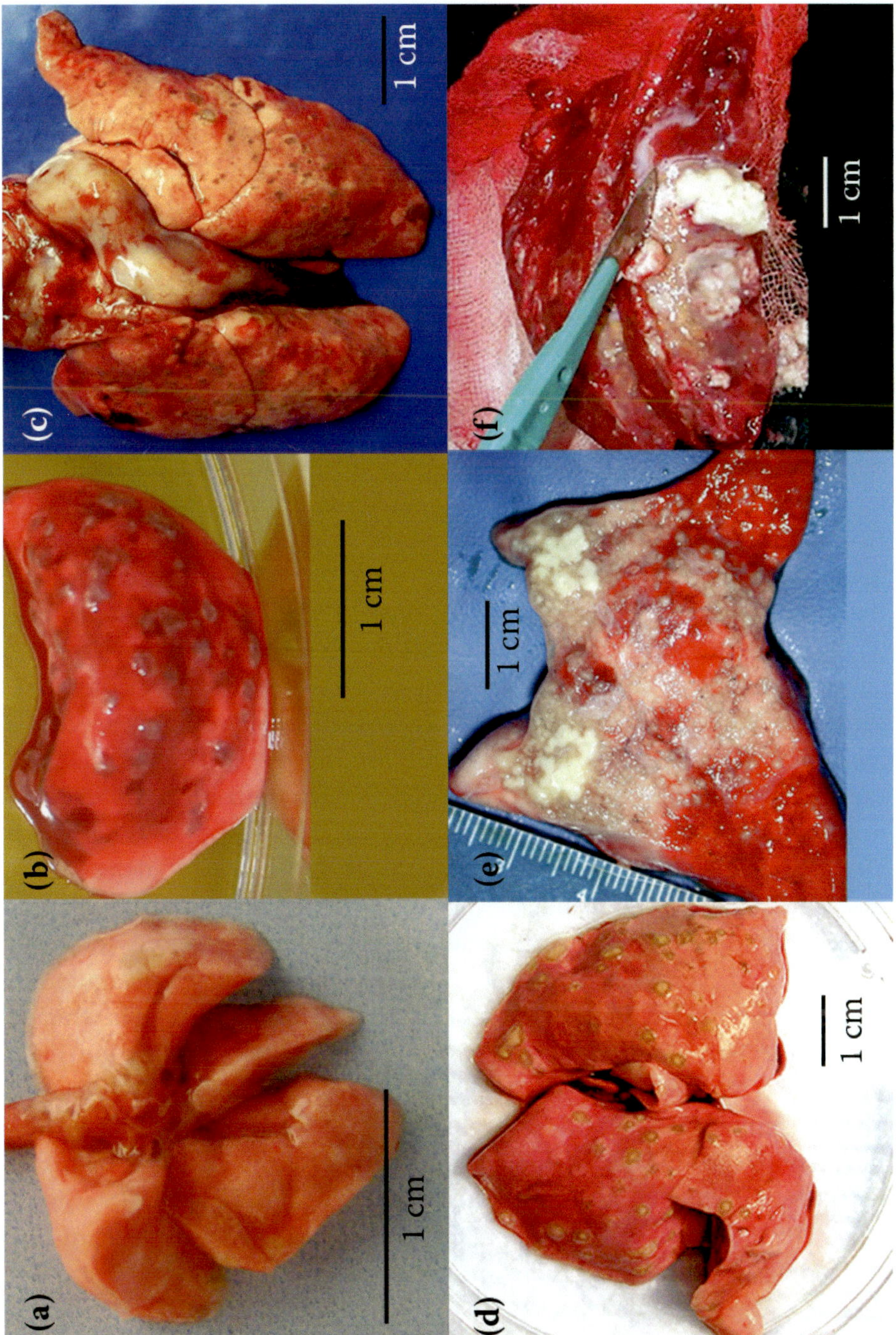
(a)
1 cm
(b)
1 cm
(c)
1 cm
(d)
1 cm
(e)
1 cm
(f)
1 cm

SUMMARY AND LESSONS LEARNED

While granulomatous inflammation characterizes the most fundamental host response to MTB infection in humans and animals, there are important species differences in disease progression and pulmonary lesion morphology which likely influence responses to drug therapy (Basaraba, 2008). In macaques, as in humans (see Chapter 5, Figure 5.1), experimentally infected animals either progress to primary active TB disease, or enter a stage of asymptomatic latent infection. In latently infected monkeys, as in humans, spontaneous reactivation can occur, and secondary TB is observed. Furthermore, the spectrum of lesion types seen in macaques shows striking similarities with that of humans. In all other models described in this atlas, one can usually differentiate two phases following infection with MTB: an early "acute" phase where total bacterial numbers increase exponentially, followed by a "chronic" phase where bacterial numbers remain roughly constant as a result of acquired immunity. In some cases, when specific combinations of animal/MTB strains are used, a reduction in bacterial load is observed rather than a plateau after the initial peak (e.g., in the rabbit—Chapter 7—or in the rat—Chapter 9). Whether this mimics immunological processes occurring in human latency remains to be determined. Spontaneous reactivation leading to secondary TB has not been observed in any of the nonprimate species. Histologically, discrete necrotizing granulomas develop from primary infection in the rabbit and guinea pig, sometimes progressing all the way to cavitation, though fibrosis is generally less pronounced than in human. Healing with fibrosis and scarring has been demonstrated, along with dystrophic calcification in older lesions. In murinae species, the pathology predominantly consists of interstitial inflammation with poorly formed nonnecrotizing granulomas. Severe infection can result in tuberculous pneumonia with areas of necrosis. Most importantly, granulomas which progress to necrosis, fibrosis, and cavitation do not occur in the murine models. Yet cavities enable spread of the disease, correlate with poor prognosis in human TB, and are suspected to constitute the most difficult to treat lesions, partly because they are likely to harbor phenotypically persistent bacilli which have become resistant to most anti-TB agents (Stewart et al., 2003). This summary of observed patterns and lesion spectrum in the various animal models is obviously a gross oversimplification because disease progression depends on many experimental factors. Inoculum size, route of infection, and timing of histopathological analyses do have major effects on the observed pathology, beside the animal/MTB strain combination. In addition to the issue of lesion diversity in terms of microbial populations and hostile microenvironments, lesion size and structure may affect drug access. Inevitably, smaller animals exhibit smaller and more homogenous granulomas, allowing for easier drug penetration compared to large fibrotic, necrotic, or even calcified human lesions which have gone through multiple cycles of healing and regrowth.

One purpose of this atlas is to help understand the differences in immunopathogenesis of experimental tuberculosis infections, from mice to nonhuman primates, and compare these observations to the situation seen in human TB. This in turn may aid in selecting the most appropriate animal models to test drugs that have been rationally designed to have specific mechanisms of action in vivo.

REFERENCES

Aber, V.R., and Nunn, A.J. 1978. Short term chemotherapy of tuberculosis. Factors affecting relapse following short term chemotherapy. *Bull Int Union Tuberc,* 53, 276–280.

Allison, M.J., Zappasodi, P., and Lurie, M.B. 1962. Host–parasite relationships in natively resistant and susceptible rabbits on quantitative inhalation of tubercle bacilli. Their significance for the nature of genetic resistance. *Am Rev Respir Dis,* 85, 553–569.

Barry, C.E.R., Boshoff, H.I., Dartois, V., Dick, T., Ehrt, S., Flynn, J., Schnappinger, D., Wilkinson, R.J., and Young, D. 2009. The spectrum of latent tuberculosis: Rethinking the biology and intervention strategies. *Nat Rev Microbiol,* 7, 845–855.

Basaraba, R.J. 2008. Experimental tuberculosis: The role of comparative pathology in the discovery of improved tuberculosis treatment strategies. *Tuberculosis (Edinb),* 88 Suppl 1, S35–47.

Boshoff, H.I., and Barry, C.E., 3rd. 2005. Tuberculosis—Metabolism and respiration in the absence of growth. *Nat Rev Microbiol,* 3, 70–80.

Chang, K.C., Leung, C.C., Yew, W.W., Ho, S.C., and Tam, C.M. 2004. A nested case-control study on treatment-related risk factors for early relapse of tuberculosis. *Am J Respir Crit Care Med,* 170, 1124–1130.

Dannenberg, A.M., Jr. 2006. *Pathogenesis of human pulmonary tuberculosis: Insights from the rabbit model,* 1st ed. Washington, DC: ASM Press.

Elwood, R.L., Wilson, S., Blanco, J.C., Yim, K., Pletneva, L., Nikonenko, B., Samala, R., Joshi, S., Hemming, V.G., and Trucksis, M. 2007. The American cotton rat: A novel model for pulmonary tuberculosis. *Tuberculosis (Edinb),* 87, 145–154.

Lin, P.L., Pawar, S., Myers, A., Pegu, A., Fuhrman, C., Reinhart, T.A., Capuano, S.V., Klein, E., and Flynn, J.L. 2006. Early events in *Mycobacterium tuberculosis* infection in cynomolgus macaques. *Infect Immun,* 74, 3790–3803.

Lin, P.L., Rodgers, M., Smith, L., Bigbee, M., Myers, A., Bigbee, C., Chiosea, I., Capuano, S.V., Fuhrman, C., Klein, E., and Flynn, J.L. 2009. Quantitative comparison of active and latent tuberculosis in the cynomolgus macaque model. *Infect Immun,* 77, 4631–4642.

Lurie, M.B. 1949. The use of the rabbit in experimental chemotherapy of tuberculosis. *Ann N Y Acad Sci,* 52, 627–636.

McMurray, D.N. 2000. A nonhuman primate model for preclinical testing of new tuberculosis vaccines. *Clin Infect Dis,* 30 Suppl. 3, S210–212.

Nuermberger, E. 2008. Using animal models to develop new treatments for tuberculosis. *Semin Respir Crit Care Med,* 29, 542–551.

Orme, I.M. 2003. The mouse as a useful model of tuberculosis. *Tuberculosis (Edinb),* 83, 112–115.

Sinsimer, D., Huet, G., Manca, C., Tsenova, L., Koo, M.S., Kurepina, N., Kana, B., Mathema, B., Marras, S.A., Kreiswirth, B.N., Guilhot, C., and Kaplan, G. 2008. The phenolic glycolipid of *Mycobacterium tuberculosis* differentially modulates the early host cytokine response but does not in itself confer hypervirulence. *Infect Immun,* 76, 3027–3036.

Smith, D.W., and Harding, G.E. 1977. Animal model of human disease. Pulmonary tuberculosis. Animal model: Experimental airborne tuberculosis in the guinea pig. *Am J Pathol,* 89, 273–276.

Smith, N.H., Hewinson, R.G., Kremer, K., Brosch, R., and Gordon, S.V. 2009. Myths and misconceptions: The origin and evolution of *Mycobacterium tuberculosis. Nat Rev Microbiol,* 7, 537–544.

Stewart, G.R., Robertson, B.D., and Young, D.B. 2003. Tuberculosis: A problem with persistence. *Nat Rev Microbiol,* 1, 97–105.

Sugawara, I., Yamada, H., and Mizuno, S. 2004a. Pathological and immunological profiles of rat tuberculosis. *Int J Exp Pathol,* 85, 125–134.

Sugawara, I., Yamada, H., and Mizuno, S. 2004b. Pulmonary tuberculosis in spontaneously diabetic goto kakizaki rats. *Tohoku J Exp Med,* 204, 135–145.

Sugawara, I., Yamada, H., and Mizuno, S. 2006. Nude rat (F344/N-rnu) tuberculosis. *Cell Microbiol,* 8, 661–667.

Turner, O.C., Basaraba, R.J., and Orme, I.M. 2003. Immunopathogenesis of pulmonary granulomas in the guinea pig after infection with *Mycobacterium tuberculosis. Infect Immun,* 71, 864–871.

Via, L.E., Lin, P.L., Ray, S.M., Carrillo, J., Allen, S.S., Eum, S.Y., Taylor, K., Klein, E., Manjunatha, U., Gonzales, J., Lee, E.G., Park, S.K., Raleigh, J.A., Cho, S.N., McMurray, D.N., Flynn, J.L., and Barry, C.E., 3rd. 2008. Tuberculous granulomas are hypoxic in guinea pigs, rabbits, and nonhuman primates. *Infect Immun,* 76, 2333–2340.

Young, D.B., Gideon, H.P., and Wilkinson, R.J. 2009. Eliminating latent tuberculosis. *Trends Microbiol,* 17, 183–188.

II

CHAPTER 4

Anatomy and Histology of the Human Lung

F. Joel W.-M. Leong and Anthony S.-Y. Leong

CONTENTS

GROSS STRUCTURE

The anatomy and histology of the human lung are well described (Breeze and Wheeldon, 1977; Gail and Lenfant, 1983; Langston et al., 1984; Kayser, 1992; Hasleton and Curry, 1996; Colby and Yousem, 1997), and this chapter serves as a review of the basic features necessary to understand the functional and pathological changes that can occur in this organ.

Lobes and Lobules

Human lungs are divided into five lobes, the right lung having three lobes—the right upper, right middle, and right lower lobes—and the left lung having two lobes—the left upper and left lower lobes. In men, the average weight of both lobes is about 850 g, and in women, 750 g. The proximal bronchial branches divide these lobes into 10 bronchopulmonary segments. These segments are made up of lobules which represent the smallest anatomical compartment of the lung that is macroscopically visible. Lobules are 1 to 2 cm in diameter, polygonal, and bound by complete or incomplete connective tissue interlobular septa. They are visible immediately beneath the pleura when outlined by septal lymphatics, especially when the latter contain anthracotic pigment or soot as occurs in smokers (Figure 4.1). Lobules are not well defined in the central regions of the lung because of age-related changes. Each lobule is made up of 3 to 30 acini and is served by centrally located terminal bronchiole–arterial bundles with pulmonary veins transporting blood away in the interlobular septae (Figure 4.2). The visualization of pulmonary lobules has importance in the correlation of radiological findings, especially from high-resolution computed tomography (e.g., interlobular septae are widened and accentuated in pulmonary edema, becoming visible radiologically as Kerley B lines).

The continued use of the term "lobule" can be confusing, because in many ways this term has been superseded by the term "acinus" as the basic practical unit of lung anatomy. However, some diseases such as emphysema are still classified according to the lobular distribution of the pathology, although the definition is based on involvement of the respiratory bronchiole and not the terminal bronchiole. For example, centrilobular (centroacinar) emphysema occurs when the walls of respiratory bronchioles are destroyed, but the more distal alveolar duct and alveoli remain intact, and in panacinar (panlobular) emphysema there is destruction of the walls of respiratory bronchioles, alveolar ducts, and alveoli.

Acini

Pulmonary acini represent the functional unit of the lung where gas exchange takes place (Raskin, 1982). They are not visible as defined units, either grossly or microscopically, and have been delineated by corrosion casts. Acini can be defined as the complex of all airways that are distal to

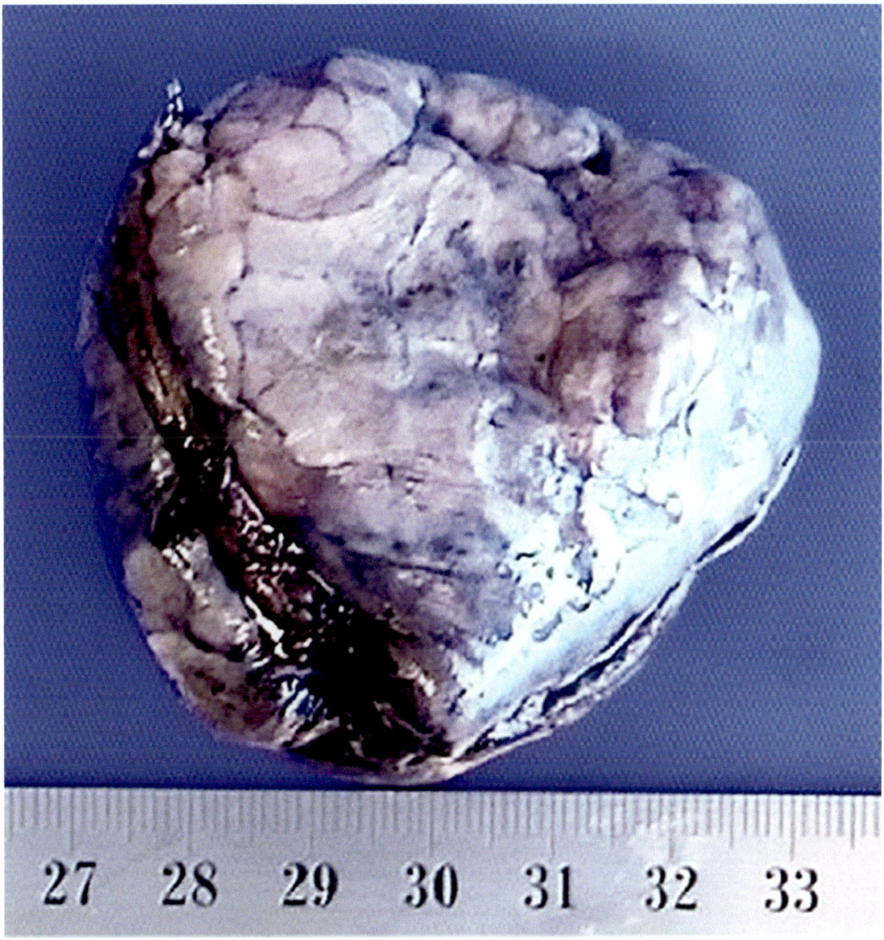

FIGURE 4.1 Inflated wedge excision of lung. Formalin-inflated wedge excision of lung showing lobular architecture and anthracosis.

terminal bronchioles. Acini thus include multiple respiratory bronchioles and their corresponding alveolar ducts, alveolar sacs, and alveoli. Each acinus averages 187 mm^3 and may be up to 9 mm in greatest diameter. There are about 25,000 acini in normal adult male lungs, with a total volume of 5.25 liters. Acini have no septal boundaries, so that collateral ventilation can occur. They are histologically important in defining the intraparenchymal spread of tuberculosis and in the classification of emphysema into centriacinar and panacinar variants. Because bronchioles are localized in the center of the lobule, the terms panacinar and

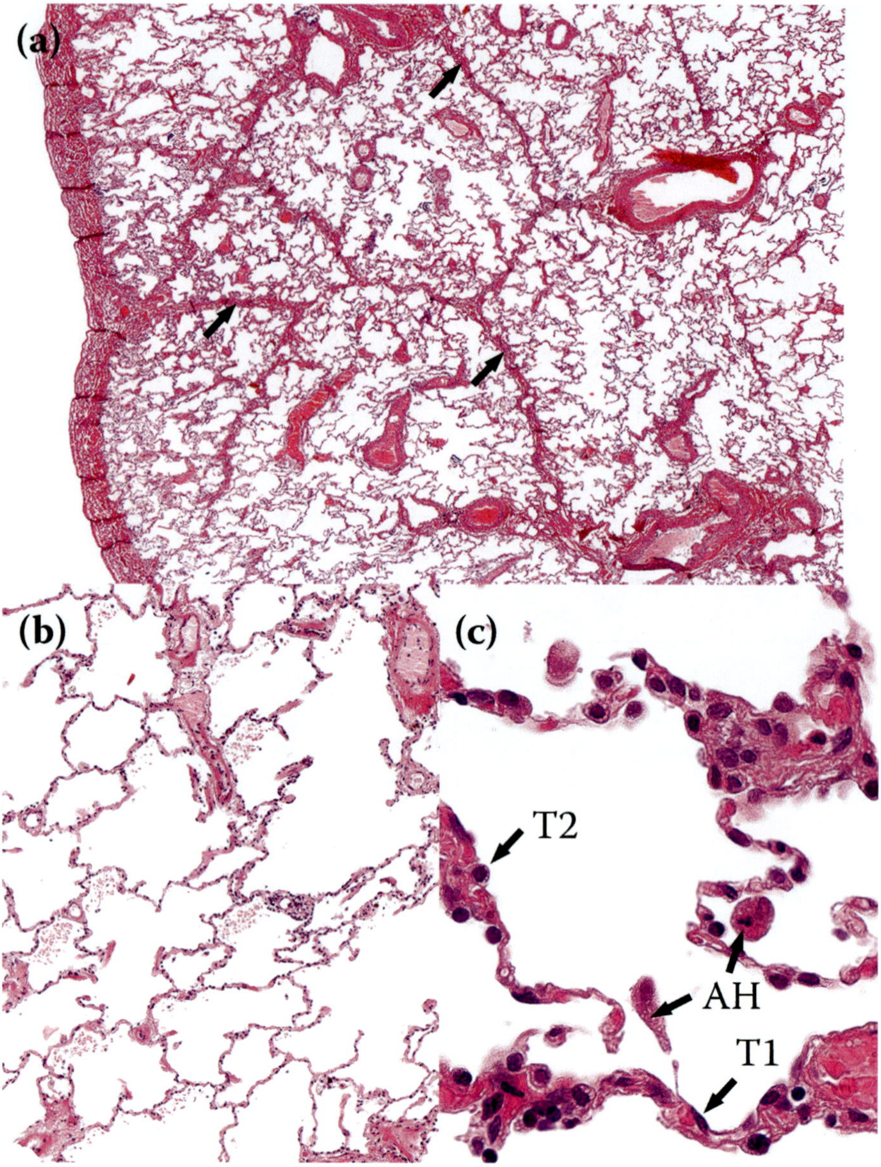
(a)
(b)
(c)
T2
AH
T1

FIGURE 4.2 (opposite page) Normal lung, hematoxylin and eosin (H&E) stain. Microwave-irradiated lung perfused with formalin. (a) LP. Thin fibrous septae (arrows) separate the lung parenchyma into distinct lobules. (b) MP. Alveolar spaces will only be well defined if the lung has been inflated prior to sectioning. (c) At high power Type I alveolar pneumocytes (T1) have central flat nuclei and broad cytoplasm, allowing them to cover 90% of the alveolar surface area. Type II pneumocytes (T2) are cuboidal cells with a central nucleus and prominent nucleolus. They are more common in number, but only cover about 5% of the alveolar surface and secrete surfactant. The expanded air spaces are empty and contain only a few macrophages/histiocytes or desquamated pneumocytes. Under normal conditions, inflammatory infiltrates, especially eosinophils are absent. AH—alveolar histiocyte.

centriacinar have been used synonymously with panlobular and centrilobular, respectively (*see above*).

Alveolar Sacs and Alveoli

The distal unit of the lung is formed by multifaceted and cup-shaped compartments known as alveoli. Where bronchiolar epithelium is completely replaced by alveolar cells, the air passage is known as the alveolar duct, and this terminates in a semicircular blind end called the alveolar sac, which is surrounded by four or more alveoli. Alveoli have mean diameters of 250 μm, and in the average male there are about 300 million alveoli with a gas-exchanging alveolar surface of approximately 143 m^2.

Progressive dilatation of air spaces occur after the age of 30 or 40 years, and alveolar ducts enlarge (alveolar duct ectasia) while adjacent alveoli appear flattened, although it is uncertain if there is actual destruction of alveolar septae. These changes are recognized as "aging lung" changes rather than the previous term of "senile emphysema."

MICROSCOPIC STRUCTURE

Diagnostic Techniques

A great deal of information can be obtained by examination of properly prepared sections of lung tissue fixed by perfusing 10% buffered formalin through the bronchial tree for at least 8 hours, or by irradiating with microwaves following formalin perfusion, or by immersion in formalin and microwave irradiated for 10 minutes at 70°C. The morphological appearances of various pulmonary structures can be further enhanced with histochemical and immunohistological stains. Histochemical stains include reticulin which displays the reticulin network, Verhoeff van Gieson stain that demonstrates collagen and elastic fibers, Mallory's trichrome for collagen and muscle, and periodic-acid-Schiff stain for mucins. Immunohistological stains can be performed with antibodies directed to smooth muscle actin, cytokeratins, surfactant apoprotein, chromogranin, and synaptophysin for neuroendocrine cells and thyroid transcription factor-1 (TTF-1), which is expressed by both bronchial epithelium and alveolar cells. Some examples of these stains are shown in the figures that follow.

Trachea and Bronchi

The conducting airways are made up mainly of smooth muscle, connective tissue, and epithelium. There are 16 to 20 rings of cartilage that encircle the anterior and lateral walls of the trachea; the posterior wall remains membranous to allow food boluses traveling down the adjacent posteriorly located esophagus to expand against the trachea. These rings of hyaline cartilage can calcify and ossify with age. The noncartilaginous parts of the trachea are composed of a membrane of collagen and elastic fibers, the latter allowing recoil from stretching during breathing.

In the main bronchi just proximal to the lobar bronchi, the cartilage is arranged circumferentially into irregular plates that decrease in size and density as they extend into the lung. With this decrease in cartilage mass is a concomitant increase in circular smooth muscle bundles to become the most prominent component in the wall of medium-sized bronchi, accounting for the efficient bronchoconstriction that occurs at this site. Some longitudinal bundles of smooth muscle, external to the circular muscle bundles, may also be present (Figure 4.3).

The amount of smooth muscle diminishes as the airways progress peripherally, and muscle is seldom found in the walls of normal terminal bronchioles. Smooth muscle proliferates centrifugally in chronic obstructive airways disease and can be prominent even around the orifice of alveolar ducts.

Elastic fibers are arranged longitudinally in the larger bronchi and become helical toward the peripheral airways, like coiled metal springs, extending into the alveolar walls as a fine reticular network (Figure 4.4).

The mucosal lining of the trachea and bronchi is continuous with the larynx and is a pseudostratified ciliated columnar epithelium, with interspersed goblet cells resting on a distinct basement membrane. The latter is mainly composed of type IV collagen, fibronectin, laminin,

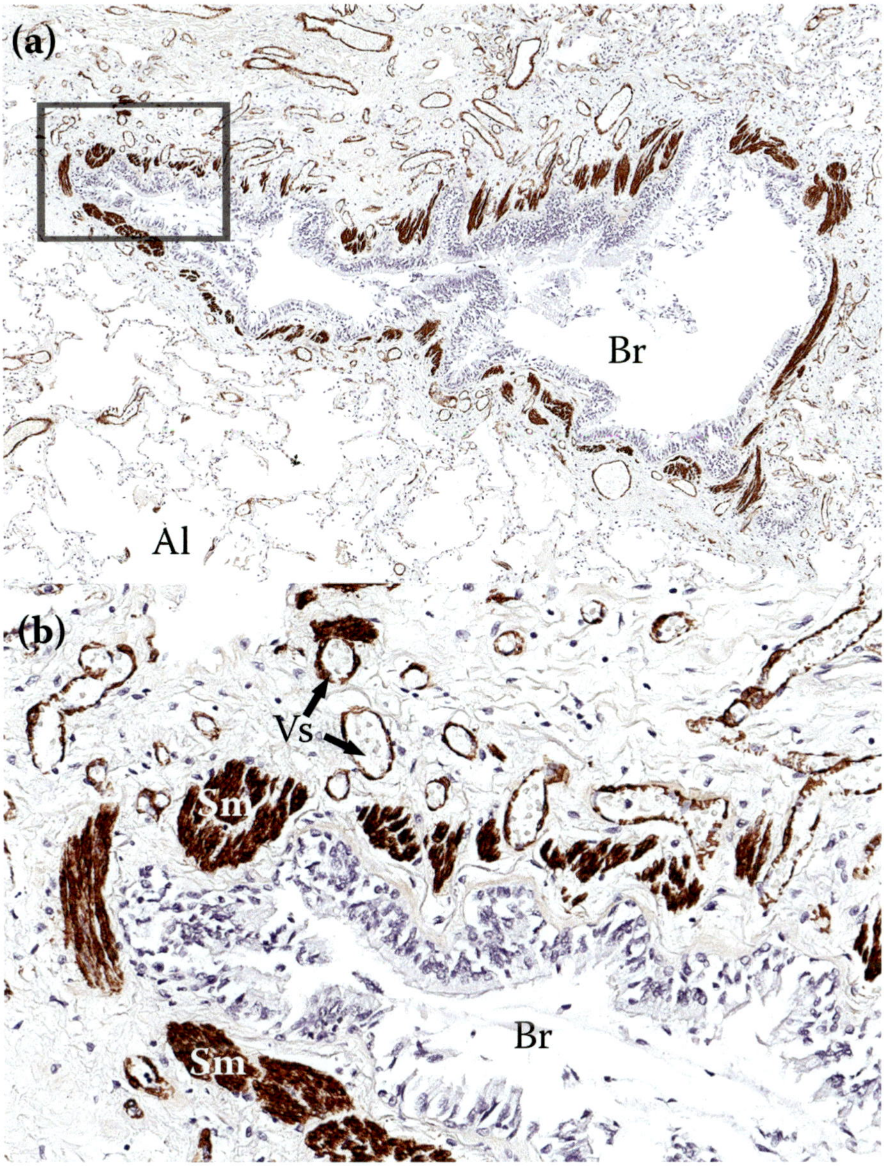

FIGURE 4.3 Normal lung smooth muscle. (a) LP. Muscle bundles in the bronchus are highlighted by staining for smooth muscle actin. (b) HP. There is also labeling of smooth muscle fibers in the adjacent blood vessels (Vs). Numerous vascular and lymphatic vessels in the bronchial submucosa, the somewhat thickened hyaline basal membrane, and a few mononuclear inflammatory infiltrates are all within normal findings. Br—bronchiole; Sm—smooth muscle; Al—alveoli.

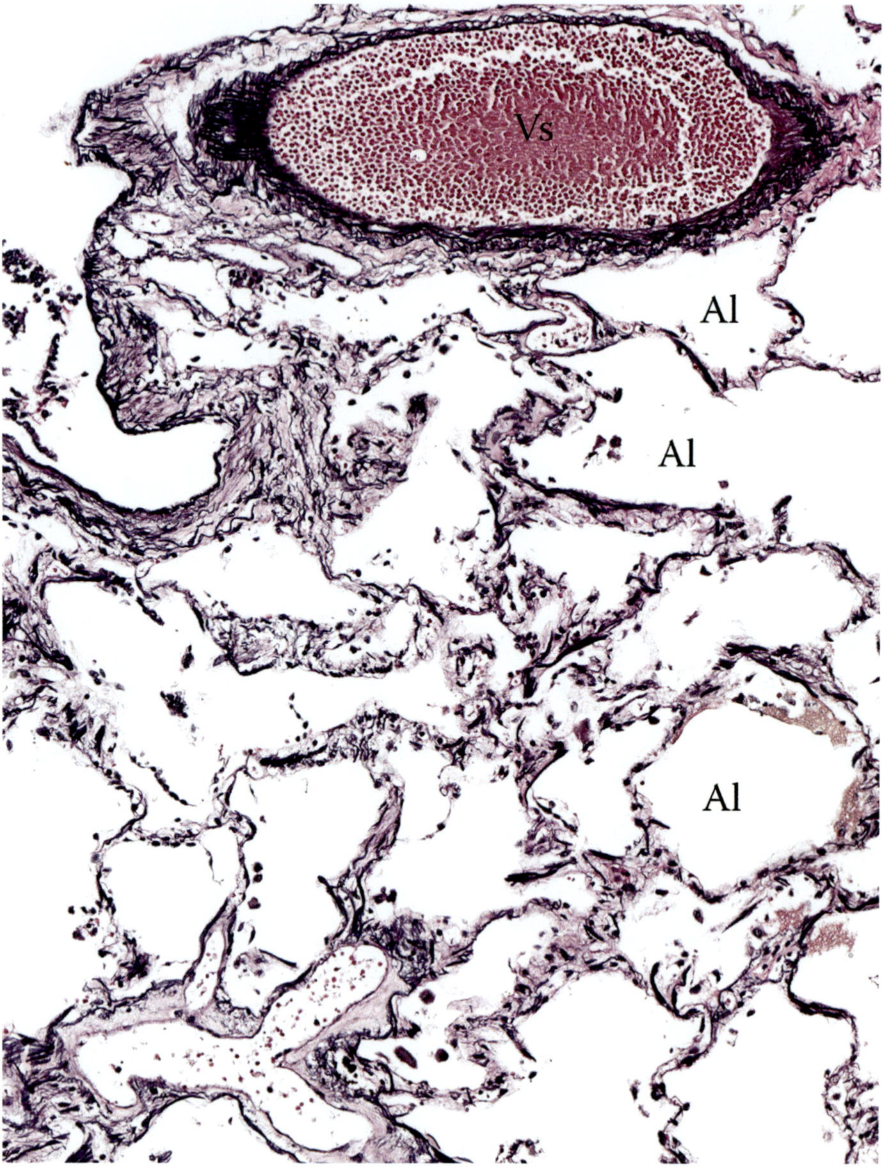

FIGURE 4.4 Normal lung elastin fibers (MP). Verhoeff van Gieson stain highlights the abundant elastic fibers that are present in the alveolar walls. The peribronchial and perivascular elastic fibers form a dense and flexible network that is often replaced by fibrous, nonelastic tissue in chronic infectious and inflammatory diseases; Al—alveoli; Vs—vessel.

and entactin–nidogen complexes, and proteoglycans (Laitinen and Laitinen, 1994). It comprises a bilaminar structure made up of the basal lamina (the true basement membrane) and a deeper lamina reticularis. Thickening and fibrosis of the lamina reticularis is characteristic of bronchial asthma. The lamina propria beneath the basement membrane contains elastic fibers and the submucosa displays mucus and serous glands (Kayser, 1992).

Four major cell types are recognized in the mucosa, namely, ciliated cells, goblet cells, basal cells, and neuroendocrine cells. The mucosa is covered by the airway surface liquid, that together with the cilia, makes up the mucociliary escalator that traps foreign particles and organisms that enter the airways, propelling them up and out of the airways via the larynx (Ng et al., 2004).

Ciliated Cells

The primary function of the columnar ciliated cells is to propel foreign particles and organisms via the mucociliary clearance mechanism. Cilia appear to cover the entire surface of the trachea, bronchi, and bronchioles, the size of the ciliated cell and length of cilia decreasing with decreasing bronchial diameter (Figure 4.5). Cilia extend into the overlying liquid layer, which has a low-viscosity or watery lower layer produced by the serous mucous glands in the submucosa and a high-viscosity or gel upper layer secreted by the goblet cells in the epithelium (Figure 4.5). The beating of the cilia propels the gel or viscous mucus on top of the low-viscosity layer to the larynx where it can be expectorated (Rutland et al., 1982).

Goblet Cells

Goblet cells are the most common nonciliated cell and are present in all bronchi but not in the bronchioles. There are as many as 7 to 25 ciliated cells for every goblet cell, the largest ratio seen in the large bronchi (Figure 4.5). The goblet cell is characterized by abundant apical collections of mucus granules that form the "goblet" appearance. These granules are released into the bronchial lumen by fusion and fenestration of the membranes of both granules and cell, in the manner of merocrine-type secretions. Secretion of mucus in the lung, as in the intestines, is thought to serve a protective function, but excess secretion may obliterate the air passage, especially in small scarred airways.

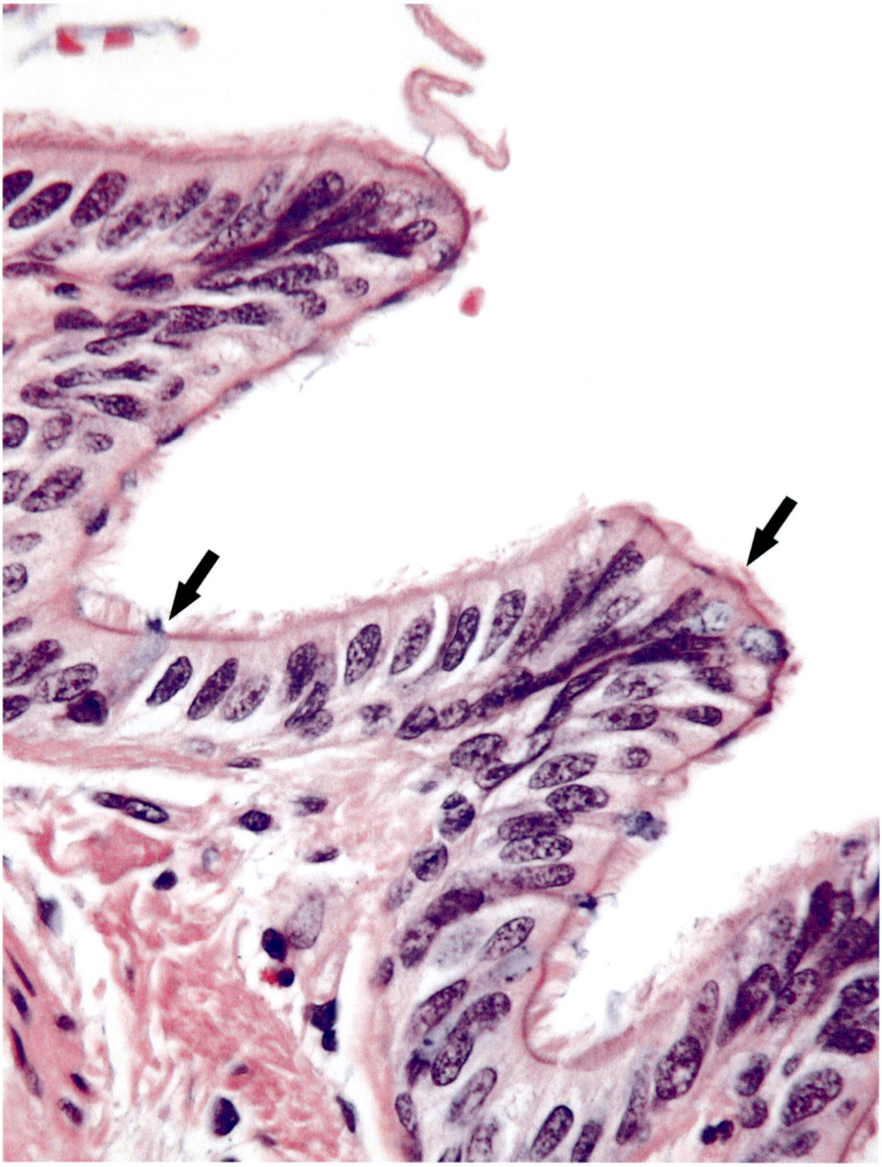

FIGURE 4.5 Normal bronchial epithelium (H&E, HP). Bronchial epithelium showing a columnar epithelium displaying a dynamic, but constant cell population composed of ciliated cells and very occasional goblet cells (arrows) at the inner surface and multipotent basal cells attached to the basal membrane.

Basal Cells

The basal cell is the progenitor of all ciliated and nonciliated cells in the bronchial mucosa. This pluripotential cell usually survives injuries that affect other mucosal cells, thus ensuring the potential to completely replenish the bronchial epithelium. Prolonged irritation stimulates proliferation of basal cells, or so-called basal cell or reserve cell hyperplasia, or induces

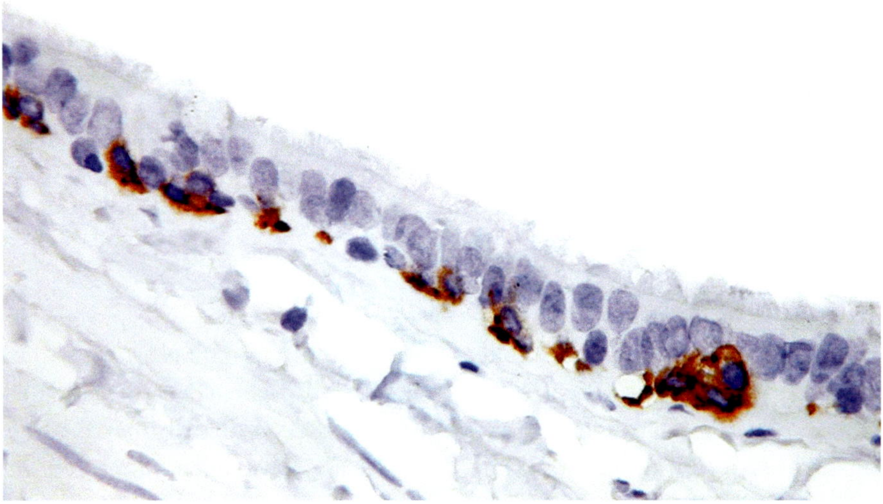

FIGURE 4.6 Normal bronchial epithelium (chromogranin, HP). Neuroendocrine cells visualized by positive staining for chromogranin are present among the basal cells and form part of the cellular composition of the bronchial epithelium.

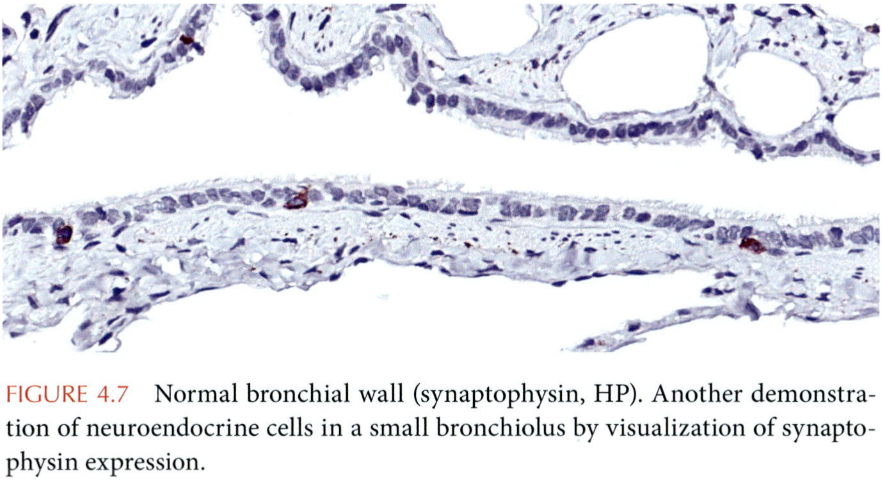

FIGURE 4.7 Normal bronchial wall (synaptophysin, HP). Another demonstration of neuroendocrine cells in a small bronchiolus by visualization of synaptophysin expression.

their differentiation into one or more differentiated forms. Deranged or mixed cell types that contain organelles of different types of cells are common in dysplastic and neoplastic epithelium that derive from basal cells.

Neuroendocrine Cells

Neuroendocrine cells are sparse in the adult lung and are more frequent in the fetal lung. They are present as individual cells among the other cell types in the mucosa or as small groups, especially near airway bifurcations in bronchi and bronchioles. These basal-oriented cells have numerous dense core granules, identified ultrastructurally. Their specific function is not well known, and they are thought to be part of the diffuse neuroendocrine system (Figure 4.6 and Figure 4.7). Neuroendocrine cells undergo hyperplasia in nonneoplastic conditions such as bronchial obstruction, broncho-pulmonary dysplasia, eosinophilic granuloma, and pulmonary hypertensive disease.

Other Bronchial Lining Cells

Brush cells, also known as tuft, caveolated, multivesicular, and fibrillo-vesicular cells, are of obscure function. These special types of cells are characterized by blunt microvilli and the presence of disk- or rod-like inclusions of unknown function. Brush cells have been identified in mucosa from the nose to the alveoli in many species but are yet to be described in humans, except in some diseases (Reid et al., 2005).

Bronchioles

A bronchus becomes a membranous bronchiole when cartilage is completely absent. Bronchioles are 1 mm or less in diameter, and the terminal membranous bronchiole leads into the acinus, which is the functional unit of the lung. There are approximately 30,000 terminal bronchioles in the lung, and each serves about 10,000 alveoli (Miller, 1947).

Membranous bronchioles are completely lined by epithelial cells comprising ciliated columnar cells and nonciliated Clara cells. Mechanical support is derived from the tethering effect of attached elastic fibers of the surrounding alveoli. These alveolar elastic fibers connect to the adventitia of the small airways and help prevent collapse of the small airways during the final phases of expiration. Destruction in emphysematous lungs results in premature collapse of small airways and obstruction of airflow.

The terminal bronchiole gives rise to one to five generations of respiratory bronchioles which, in turn, connect with alveoli ducts, and to alveolar sacs and alveoli. The respiratory bronchiole is lined by mostly nonciliated cuboidal Clara cells and a few ciliated cells. Goblet cells are no longer present. Ciliated cells extend to the bronchiolo-alveolar junction, gradually decreasing in number as they extend peripherally and maintaining the ciliary flow around the nonciliated Clara cells. There are direct communications between the lumen of membranous bronchioles and adjacent alveoli through Lambert's canals. These are difficult to observe microscopically, but they have a definite role in collateral ventilation. Resolution of inflammatory changes may result in metaplastic growth of bronchiolar epithelium through Lambert's canals. Depressions and fistulous perforations of the wall of small membranous bronchioles also commonly occur in lung damage, especially in smokers, causing peribronchiolar proliferations and a similar metaplastic change.

The Clara cell is a nonciliated cuboidal cell found in the membranous and respiratory bronchioles. It replaces the diminishing goblet cell population in small bronchioles. The Clara cell is dome shaped and rich with mitochondria and endoplasmic reticulum. It contains a prominent Golgi, and the apical cytoplasm contains numerous periodic-acid-Schiff (PAS)-diastase positive secretory granules (Figure 4.8). Clara cells also secrete surfactant apoproteins that decrease the surface tension and keep the small air channels open. They are also thought to regulate the transport of chloride ions. These cells serve as the reserve and reparatory cell in the small airways, similar to the role of basal cell more proximally and alveolar Type II cell more distally (Jeffery et al., 1992).

Acini and Alveoli

Several alveolar ducts result from the division of a single respiratory bronchiole. The alveolar ducts terminate in alveolar sacs, which are formed by four or more alveoli (Figure 4.9). The orifices of alveoli along the alveolar ducts and alveolar sacs are formed by thick elastic and collagen bundles that continue from bronchial and bronchiolar elastic bundles. From the orifice, finer elastic bundles extend out further and are interwoven with fibers of capillaries, resembling a basket weave. This network of elastic fibers is interconnected in all directions and functions as an integrated network that is fundamental to the uniform expansion and retraction of the lung during respiration. The fine network of elastic fibers and capillaries is plated on both sides by a layer of alveolar lining epithelial cells. This very thin membrane between the air–blood interface allows gas exchange to take place. The membrane is made up of the fused basement membranes of the overlying

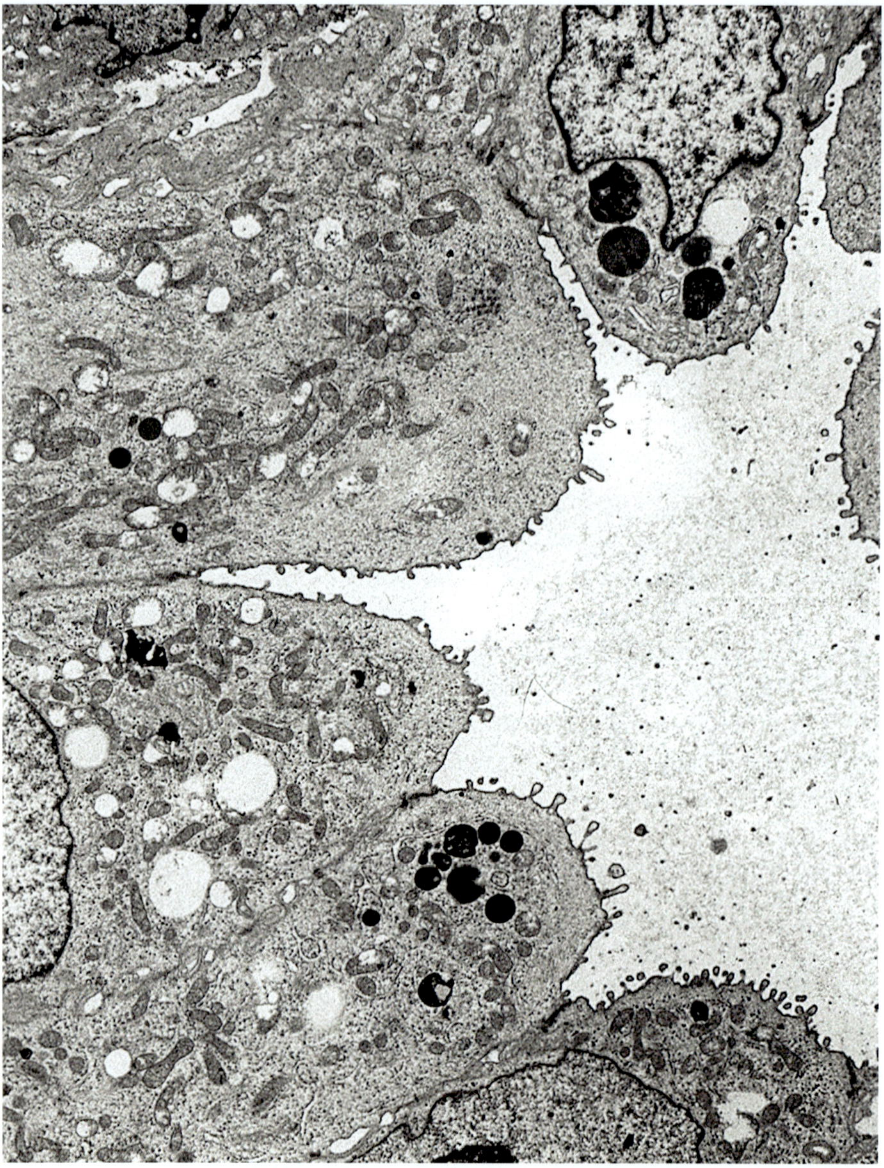

FIGURE 4.8 Clara cells (electron microscopy). Electron micrograph of Clara cells, which are dome shaped and show stubby microvillus projections on their luminal surface. The apical cytoplasm is rich in secretory granules, endoplasmic reticulum, and mitochondria.

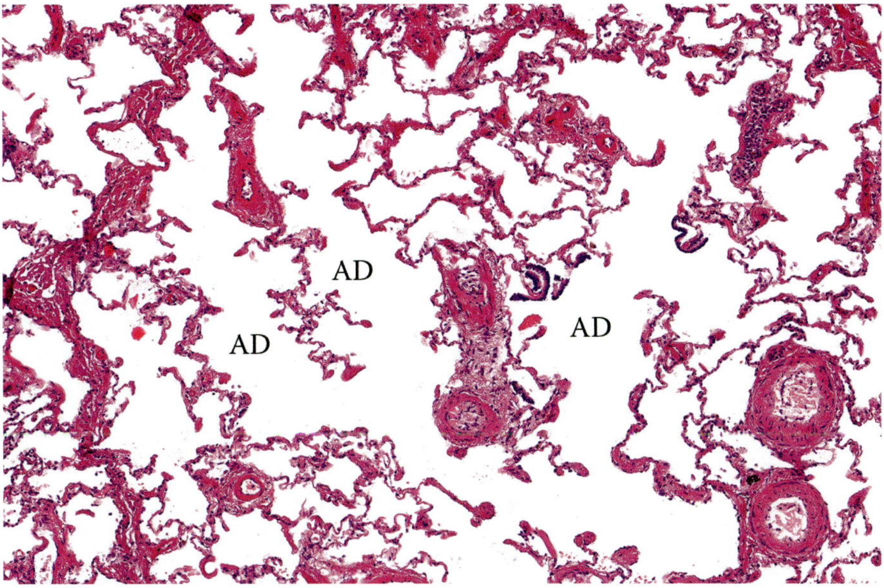

FIGURE 4.9 Normal alveolar ducts (H&E, LP). Several converging alveolar ducts are present in the center of the field in this sample of microwave-irradiated formalin-perfused postmortem lung; AD—alveolar ducts.

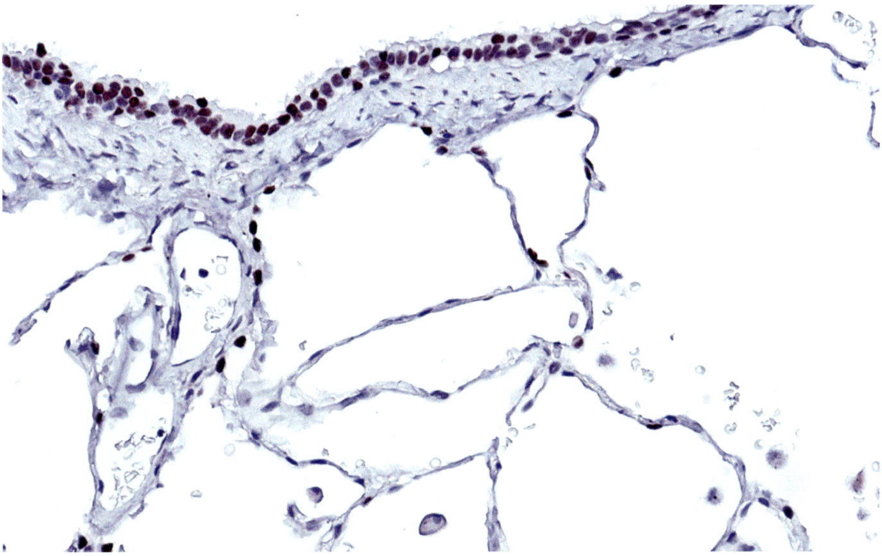

FIGURE 4.10 Bronchial and alveolar cells (thyroid transcription factor, HP). Normal bronchiolus and adjacent alveoli expressing thyroid transcription factor (TTF-1) in both the bronchial and alveolar cells.

alveolar Type I cell and the adjacent endothelial cell, with the attenuated cytoplasm of these two cells on either side. The pores of Kohn allow communication between adjacent alveoli, with possibly 13 to 21 such pores per alveolus to allow for collateral ventilation (Renkin, 1992).

Alveolar Lining Cells

There are two types of alveolar lining cells. The alveolar Type I pneumocyte has a central flattened nucleus and a broad expanse of cytoplasm that extends up to 50 μm in diameter and is as thin as 0.1 μm. Type I cells make up about 40% of the alveolar lining cells but cover 90% of the alveolar surface. These cells have well-developed basement membranes but show sparse surface microvilli and cytoplasmic organelles. Both alveolar and bronchial cells express thyroid transcription factor-1 (TTF-1), a marker useful for the identification of tumor of pulmonary origin (Figure 4.10).

Alveolar Type II pneumocytes, in contrast, are cuboidal cells with a diameter of up to 15 μm, display a large basal nucleus with a prominent nucleolus, and abundant osmophilic laminar inclusions, the precursors of surfactant. Type II cells constitute 60% of the alveolar cells but cover only about 5% of the alveolar surface (Figure 4.11). At ultrastructural level many stubby microvilli can be seen projecting into the alveolar space, and the cells have well-developed endoplasmic reticulum and Golgi apparatus. They are connected by well-formed tight junctions to adjacent Type I cells and lie on a well-defined basement membrane. Besides secreting surfactant, Type II cells serve as the reserve cell, normally maturing into the Type I cell. Type II cells are more resistant to injury and undergo hyperplasia in response to damage, and may also show dysplastic change.

When secreted into the alveolar space, the osmophilic lamellar inclusions, a partially crystallized form of tubular myelin produced in Type II pneumocytes, spread out as a thin layer of surfactant (Figure 4.12). Surfactant is composed mainly of phospholipids, especially dipalmitoyl lecithin, and glycoproteins. With deflation of the alveolus, the phospholipids are compressed and align into a layer with hydrophilic and hydrophobic ends on each side at the air–liquid interface. This arrangement reduces the surface tension and prevents the alveolus from collapsing. With inflation, this arrangement in the alveolus is disrupted, and the corresponding increase in surface tension assists the elastic recoil of the alveolus in expiration. In prematurity, insufficient production of surfactant results in collapse of alveoli and pulmonary edema as part of the hyaline membrane disease. In the diffuse alveolar damage syndrome, leakage of fibrin and

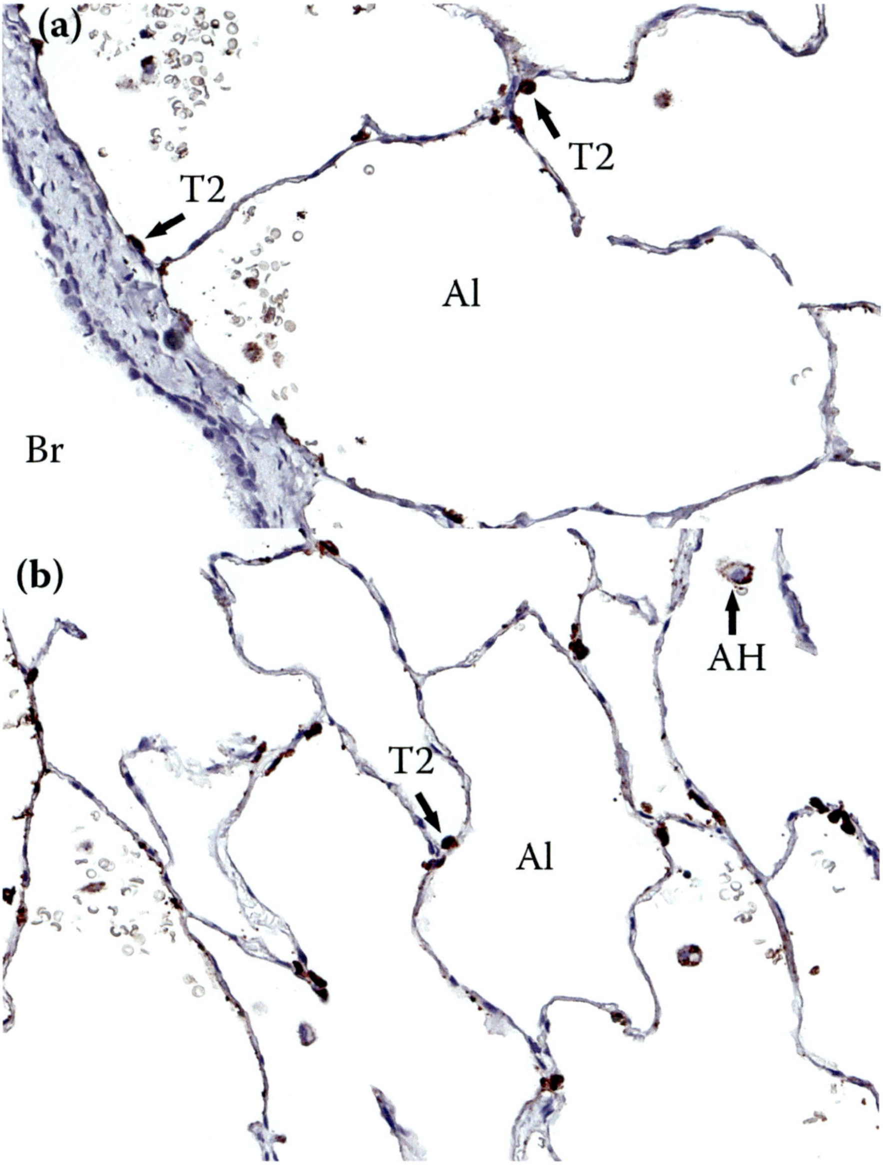

FIGURE 4.11 Alveolar cells (surfactant immunostaining, HP). (a, b) Terminal bronchiolus, adjacent alveolar ducts, and alveoli. The alveolar cells express surfactant in contrast to the columnar cells of the terminal bronchiole (Br). T2—Type II pneumocyte. AH—alveolar histiocytes. Al—alveolar sac/alveolus.

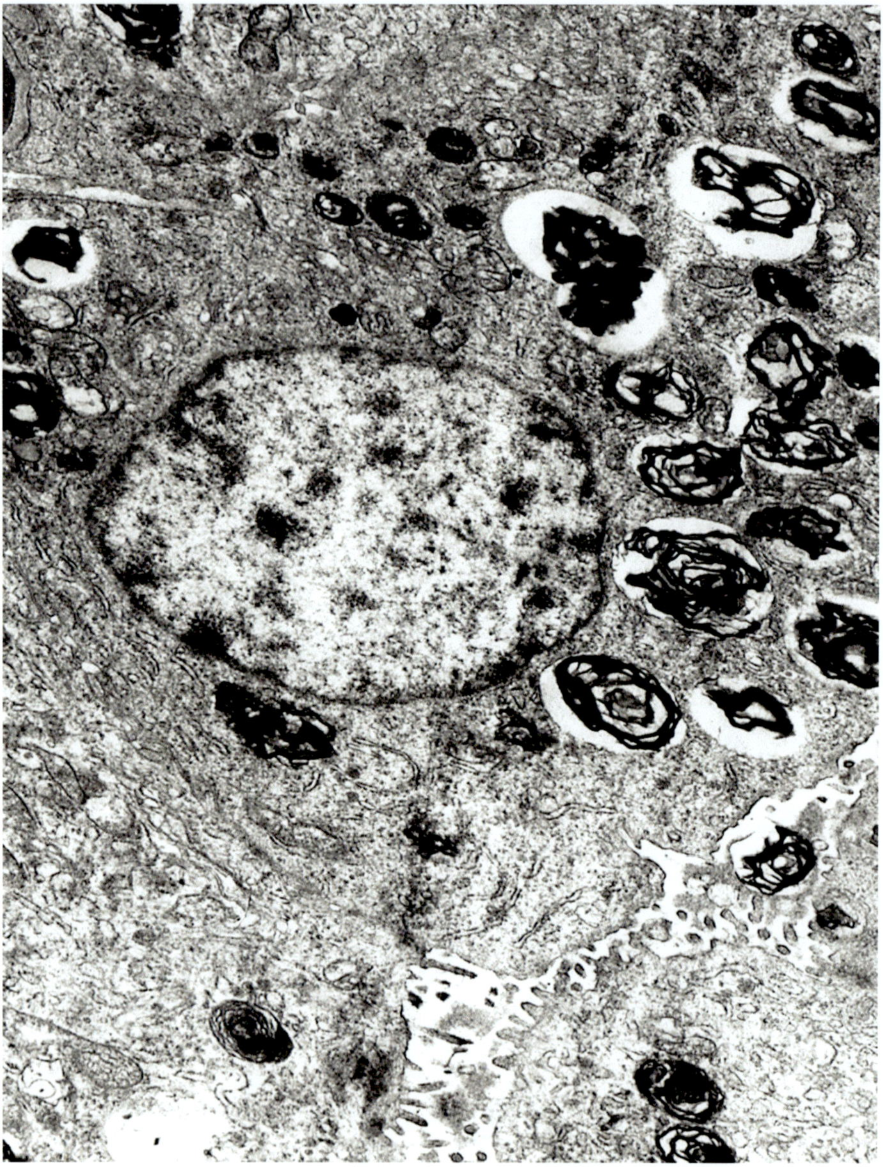

FIGURE 4.12 Type II pneumocytes. Electron micrograph of an alveolar Type II pneumocyte. These cells are characterized by the presence of numerous lamellar bodies that represent newly synthesized pulmonary surfactant which are released into the lumen. The cytoplasm is also rich in multivesicular bodies and rough endoplasmic reticulum, and microvilli are present on the cell surface.

capillary contents into the alveolus interferes with the action of surfactant (Rooney et al., 1994).

The fine structure and permeability of the alveolar capillary endothelium is similar to that of capillary endothelium elsewhere in the body. Besides serving as a barrier with active regulation of water and solute transport, alveolar endothelium also selectively processes and modifies a wide range of substances. The endothelium converts angiotensin I to angiotensin II and inactivates bradykinin by angiotensin-converting enzyme. It also clears serotonin, norepinephrine, prostaglandin E and F, adenine nucleotides, and some hormones and drugs, and releases angiotensin II, adenosine, some prostaglandins, and previously accumulated drugs and metabolites (Renkin, 1992).

Other cells in the alveolar wall include mesenchymal cells such as fibroblasts, pericytes of capillaries, and myofibroblasts that lie adjacent to the capillary in the alveolar septum. These cells are responsible for the maintenance and metabolism of the elastic and collagen fibers and proteoglycans in the alveolar walls. Collagen fibers can be seen ultrastructurally in the alveolar walls but only as delicate fibers; when apparent by light microscopy they are considered abnormal. Lymphocytes, plasma cells, neutrophils, eosinophils, basophils, and mast cells may be present in small numbers, as may be the occasional fixed or migrating macrophage. Intravascular megakaryocytes are frequently present. These appear as hyperchromatic nuclear material squeezed within capillary lumen and should not be mistaken for viral inclusions or embolic trophoblastic material. Larger numbers of neutrophils or the degranulation of neutrophils may be responsible for insidious tissue lysis, such as elastolysis in pulmonary emphysema, and increase in mast cells and eosinophils occurs in bronchial asthma or other hypersensitivity disorder.

Vascular Supply

The lung has a dual blood supply. The pulmonary circulation is a low-pressure/high-capacity system that accommodates total systemic venous return and is structurally organized to absorb a large change of the flow volume with minimal change in pressure. The bronchial arterial system serves a nutritional function and is part of the systemic circulation, delivering blood at high pressure and of high arterial oxygen content.

Histologically, the structure of the pulmonary arteries varies primarily with the vascular diameter and is of three major types: elastic, muscular, and nonmuscular. The main pulmonary artery and branches down to a diameter of about 1000 μm in the adult have an elastic structure consisting predominantly of interrupted elastic fibers arranged circumferentially, different from the uninterrupted elastic fibers seen in the aorta and in the pulmonary artery of the fetus and neonate. In persistent pulmonary hypertension that commences at birth, the continuous elastic fiber pattern of the neonate may be maintained into adulthood. Transitional arteries, roughly between 500 and 1000 μm diameter, have a mixture of elastic and muscle fibers in their walls and lead to arteries between 70 and 500 μm in external diameter. The arterial walls are of the muscular type with circumferential smooth muscle bound by a distinct internal and external elastic lamina. From a diameter of 30 to 150 μm, elastic fibers are lost, and the continuous muscle coat gives rise to a spiral of muscle. On cross section through the vessel, this appears as a crescent of muscle in the wall. Arteries of less than 70 μm diameter tend to be nonmuscular and feed directly into alveolar capillaries.

Small elastic or muscular pulmonary arteries can be difficult to discern from bronchioles, although when compared in cross section, the bronchiole will usually have a larger diameter with more irregularly orientated inner and outer elastic fibers (Kayser, 1992).

Small intra-acinar pulmonary venules have a nonmuscular wall and usually cannot be discerned from nonmuscular arterioles without the use of injection techniques. Larger pulmonary veins can be identified with elastic stains because they have a well-defined internal elastic lamina and a thin, predominantly elastic wall structure without an external elastic lamina, the mural elastic fibers merging directly with those in the adventitia. Muscle fibers are sparse and irregularly arranged. Both arteries and veins in the elderly may show nonspecific intimal hyalinized fibrosis, and in chronic pulmonary venous hypertension the veins may acquire a thick medial muscular coat and develop well-defined external elastic lamina (arterialization).

The normal bronchial artery is difficult to identify macroscopically because its diameter is only about 1.5 mm at the hilum; however, in chronic disease states such as bronchiectasis and pulmonary hypertension, the bronchial arteries hypertrophy and may become grossly visible. Histologically, it is a muscular artery with well-defined internal elastic lamina but an absent or poorly defined external elastic lamina. This contrasts with the well-defined internal and external elastic laminas of the muscular pulmonary arteries. Bronchial veins histologically resemble pulmonary veins.

The lung has a rich supply of lymphatics. The superficial or pleural plexus drains the outer parts of the lung via the visceral pleural lymphatics (pleura visceralis), and the deep or parenchymal plexus drains the bronchovascular bundles toward the hilar lymph nodes. While there is communication of the two systems at the boundaries of their distribution and between lobes or lobules and the pleura near the hilum, the two systems drain separately toward hilar nodes in large lymphatic channels that contain valves. Lymphatic vessels are not present within the interalveolar septula, within the alveolar ducts, or along the respiratory bronchioles (Kayser, 1992). Alveolar walls do not have lymphatic spaces although juxta-alveolar lymphatics in the small or distal bronchovascular bundles are partly facing and in contact with the basal surface of adjacent alveolar epithelium.

In the normal lung, lymphatics are collapsed and difficult to identify. In conditions of pulmonary edema or lymphangitis carcinomatosa, the lymphatic channels are dilated and readily observed. Lymphatic spaces are lined by large flattened endothelial cells with few organelles. Although the cells are joined by all forms of intercellular junctions, the presence of focally open or movable junctions devoid of basement membrane is a unique feature of lymphatic capillaries.

REFERENCES

Breeze, R.G., and Wheeldon, E.B. 1977. The cells of the pulmonary airways. *Am Rev Respir Dis,* 116, 705–777.

Colby, T.V., and Yousem, S.A. 1997. Lungs. In *Histology for pathologists,* ed. S.S. Sternberg. New York: Lippincott-Raven, 433–458.

Gail, D.B., and Lenfant, C.J. 1983. Cells of the lung: Biology and clinical implications. *Am Rev Respir Dis,* 127, 366–387.

Hasleton, P.S., and Curry, A. 1996. Anatomy of the lung. In *Spencer's pathology of the lung,* ed. P.S. Hasleton. New York: McGraw-Hill, 1–44.

Jeffery, P.K., Gaillard, D., and Moret, S. 1992. Human airway secretory cells during development and in mature airway epithelium. *Eur Respir J,* 5, 93–104.

Kayser, K. 1992. *Analytical lung pathology* Berlin, New York: Springer-Verlag.

Laitinen, A., and Laitinen, L.A. 1994. Airway morphology: Epithelium/basement membrane. *Am J Respir Crit Care Med,* 150, S14–17.

Langston, C., Kida, K., Reed, M., and Thurlbeck, W.M. 1984. Human lung growth in late gestation and in the neonate. *Am Rev Respir Dis,* 129, 607–613.

Miller, W.S. 1947. *The lung,* 2nd ed. Springfield, IL: Charles C Thomas.

Ng, A.W., Bidani, A., and Heming, T.A. 2004. Innate host defense of the lung: Effects of lung-lining fluid pH. *Lung,* 182, 297–317.

Raskin, S.P. 1982. The pulmonary acinus: Historical notes. *Radiology,* 144, 31–34.

Reid, L., Meyrick, B., Antony, V.B., Chang, L.Y., Crapo, J.D., and Reynolds, H.Y. 2005. The mysterious pulmonary brush cell: A cell in search of a function. *Am J Respir Crit Care Med,* 172, 136–139.

Renkin, E.M. 1992. Cellular and intercellular transport pathways in exchange vessels. *Am Rev Respir Dis,* 146, S28–31.

Rooney, S.A., Young, S.L., and Mendelson, C.R. 1994. Molecular and cellular processing of lung surfactant. *FASEB J,* 8, 957–967.

Rutland, J., Griffin, W.M., and Cole, P.J. 1982. Human ciliary beat frequency in epithelium from intrathoracic and extrathoracic airways. *Am Rev Respir Dis,* 125, 100–105.

CHAPTER 5

Pathology of Tuberculosis in the Human Lung

F. Joel W.-M. Leong, Seokyong Eum,
Laura E. Via, and Clifton E. Barry, 3rd

CONTENTS

INTRODUCTION

The histological appearances of tuberculosis in man are characterized by highly developed reactivated heterogeneous lesions that are larger than those found in most animals. Necrosis or caseation, as well as fibrosis, are prominent histological features of such lesions, and secondary complications are commonplace. Tuberculous granulomas can present in a range of sizes from fractions of a millimeter to several centimeters in the largest dimension. It is well recognized that untreated tuberculosis (TB) lesions can grow to a large size and may appear to be composed of multiple coalesced individual lesions that can cause organ dysfunction depending on their location.

Early clinical symptoms of tuberculosis include chronic cough, fever, anorexia, weight loss, and hemoptysis. Clinical examination of the chest may reveal no abnormality or may disclose signs of tachypnea, wheeze, decreased breath sounds, or bronchial breath sounds (depending on the nature of the underlying complication). TB has a range of histological appearances, and these appearances are part of a dynamic process.

The tissues presented in this chapter were collected from adult patients undergoing lung resection for the management of multidrug-resistant (MDR) TB. The subjects (both male and female) ranged from 19 to 50 years of age and had an average of 16.8 months of treatment administered in an average of 2.2 separate drug regimens prior to surgery. Both primary MDR and acquired MDR disease were represented among these subjects. Surgery was an elective procedure offered by the hospital to patients in generally good physical condition, with highly localized disease typically in one of the two upper lobes of the lung as described previously (Park et al., 2002, 2009). This tissue collection was approved by the National Masan Tuberculosis Hospital (NMTH) institutional review board and granted an exemption by the U.S. National Institutes of Health, Office of Human Subject Research. Resected lung tissue was collected with the written consent of these subjects.

All patients were treated and underwent surgery at the NMTH in Masan, Republic of Korea. The hospital is a 430-bed facility that has served as a national referral center since the 1970s for individuals with complicated, recurrent and/or drug-resistant tuberculosis. The facility has access to the full complement of antituberculous agents typically used to treat MDR-TB and a surgical department with well-qualified staff. Patients are commonly prescribed individualized combination regimens of at least five agents including a fluoroquinolone (ofloxacin, levofloxacin, or moxifloxacin), an injectable (streptomycin, amikacin, or kanamycin), and three of the following: pyrazinamide, ethambutol, prothionamide, cycloserine, and para-aminosalicylic acid for approximately two years (18 months after sputum culture conversion).

Tuberculosis is generally thought to be composed of primary TB and secondary disease. It is generally uncommon in clinical practice to encounter a primary TB lesion, unless it is an incidental finding at autopsy. It is not possible to determine whether a granuloma is primary or secondary based on histological appearances alone, and care should be taken not to apply clinical or macroscopic terms to histological interpretations.

Primary Tuberculosis

Primary TB is the infection of an individual lacking any previous history of TB and is thought to arise shortly after exposure to an infectious index case. In primary lung infection, this manifests as a single lesion (referred to as a Ghon focus) with a characteristic location—immediately subadjacent to the pleura in the lower part of the upper lobes or upper part of the lower lobes of one lung (Ghon, 1912). The localization reflects the parts of the lung receiving the greatest volume flow of inspired air. It is rare, but possible to find bilateral or multiple foci. The Ghon focus is typically a rounded, well-circumscribed gray–white consolidated inflammatory focus, 10 to 15 mm in maximum dimension. It is an evolving tuberculous granuloma that is initially solid and composed of epithelioid histiocytes intermixed with chronic inflammatory cells, but develops a soft, necrotic center within a week. *Mycobacterium tuberculosis* (MTB), either free or within phagocytes, drain along the regional peribronchial lymphatic channels to the ipsilateral tracheobronchial lymph nodes, where they can elicit more caseating granulomas. A Ghon complex is the combination of the primary lung lesion and an involved lymph node. If the initial site of infection is not the lung, similar primary complexes can form at the site of infection and at the draining lymph nodes. Less common entry points include the gastrointestinal tract, oropharyngeal lymphoid tissue, and skin.

There are several possible pathways from primary tuberculosis (Figure 5.1). Most commonly, the lesions do not progress—they undergo a degree of shrinkage and involute, as evidenced by fibrosis, calcification, and sometimes ossification. The overlying pleural surface can become puckered due to fibrous scarring. If fibrocalcific scarring replaces most of the tuberculous granuloma (Figure 5.2d), it can become quite difficult to detect with imaging and only be discovered as an incidental finding at autopsy. Healing of primary tuberculosis does not mean the bacteria are eradicated—bacilli are thought to persist in a dormant state for years, only to become active if the immune system is suppressed. Sometimes, primary tuberculosis does not heal. The lesion may enlarge, erode into the bronchial tree (Figure 5.3), and produce multiple new lesions close by (satellite lesions, Figure 5.3a, Figure 5.4). Hilar lymph nodes can also enlarge and become confluent caseous masses that compress or obstruct major

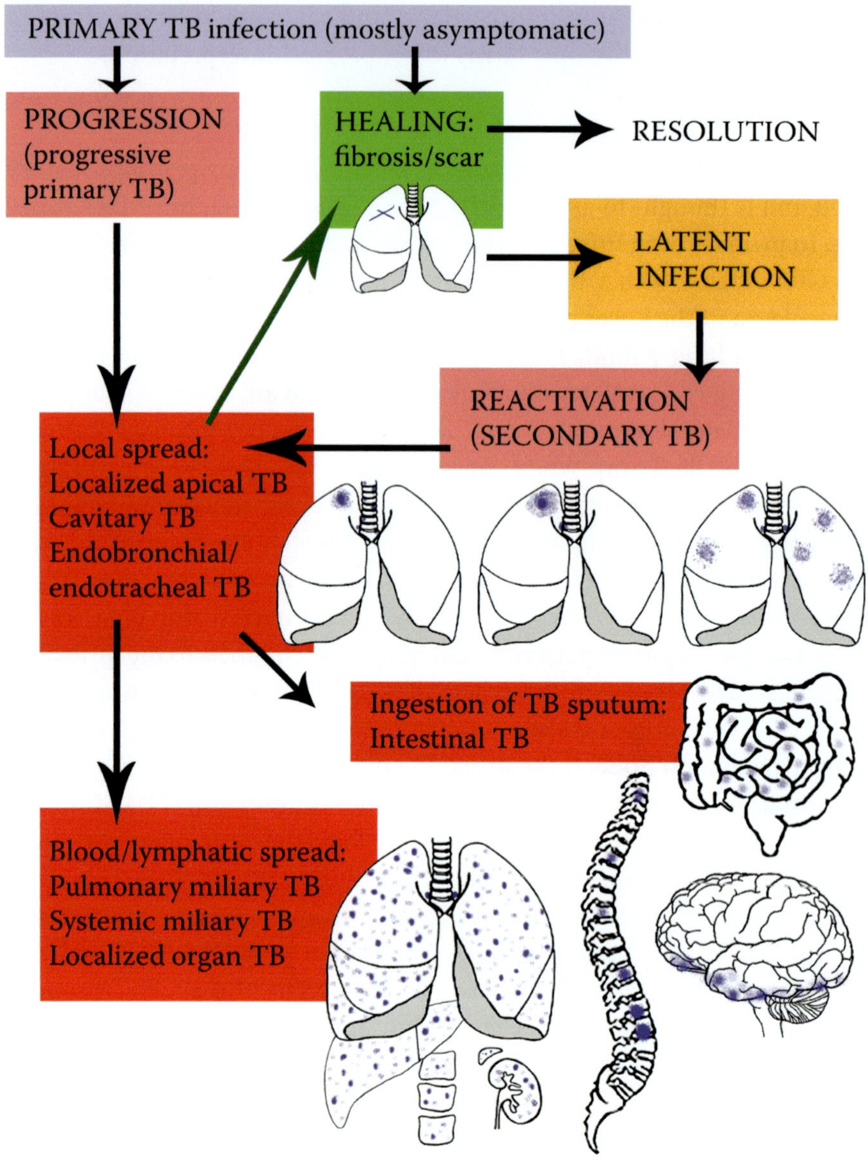
PRIMARY TB infection (mostly asymptomatic)
PROGRESSION (progressive primary TB)
HEALING: fibrosis/scar
RESOLUTION
LATENT INFECTION
REACTIVATION (SECONDARY TB)
Local spread: Localized apical TB Cavitary TB Endobronchial/ endotracheal TB
Ingestion of TB sputum: Intestinal TB
Blood/lymphatic spread: Pulmonary miliary TB Systemic miliary TB Localized organ TB

FIGURE 5.1 (opposite page) Primary TB sequelae. This diagram outlines the possible outcomes following primary TB infection. All pathways can be understood as interaction between the (slowly multiplying) infectious organisms and the defense mechanisms of the host. The infection may regress and disappear or leave a scar. Most will never know they were infected. Some, however, will hold the infection in metastable or latent form and have it reactivate years later due to some precipitant such as immunosuppression. Reactivation TB is also referred to as secondary TB. In rare cases, primary infection does not regress and heal, but progresses immediately to active disease. Both progressive primary TB and secondary TB (reactivation) can follow similar pathways. Local spread can occur by expansion of the infected site (single apical lesion ± hilar node involvement), or through spread via the airways to affect bronchi and trachea (endobronchial/endotracheal TB). Dissemination through blood or lymphatics can result in granulomas throughout the lung (pulmonary miliary TB), in major organs, in bone, brain, and meninges. If untreated, any combination is possible. Accidental swallowing of infected sputum can spread TB to the gastrointestinal tract. Early local disease can also heal and regress (mostly due to treatment but rarely spontaneously), and even return to some form of "latent" state.

bronchi. If there is vascular dissemination, miliary tuberculosis can result (Andreu et al., 2002; Kumar et al., 2010).

Secondary Tuberculosis

Secondary tuberculosis arises in an individual who has been exposed to MTB in the past. Most cases are thought to be reactivation of asymptomatic primary disease (Hunter et al., 2007). The factors which initiate this change are of great interest. There is speculation that pathogen-induced dysregulation of host lipid synthesis and sequestration plays a key role in this transition, which would elevate the foamy macrophage (Figure 5.5) from an incidental finding to a key participant in the granuloma (Russell et al., 2009). During primary infection it is thought to be possible for bacilli to disseminate without producing symptoms, establishing themselves in sites with high oxygen tension, particularly the lung apices. This can occur at any time following a primary infection, and the usual precipitant is a weakening of the immune system. Traditional teaching is that secondary TB begins in the apical or posterior segment of one or both upper lobes. This is in close proximity to the clavicle on chest X-ray films. In clinical practice, secondary tuberculosis can be found at a variety of sites around the lung, such as the hilum, and inevitably, regional lymph nodes are also affected.

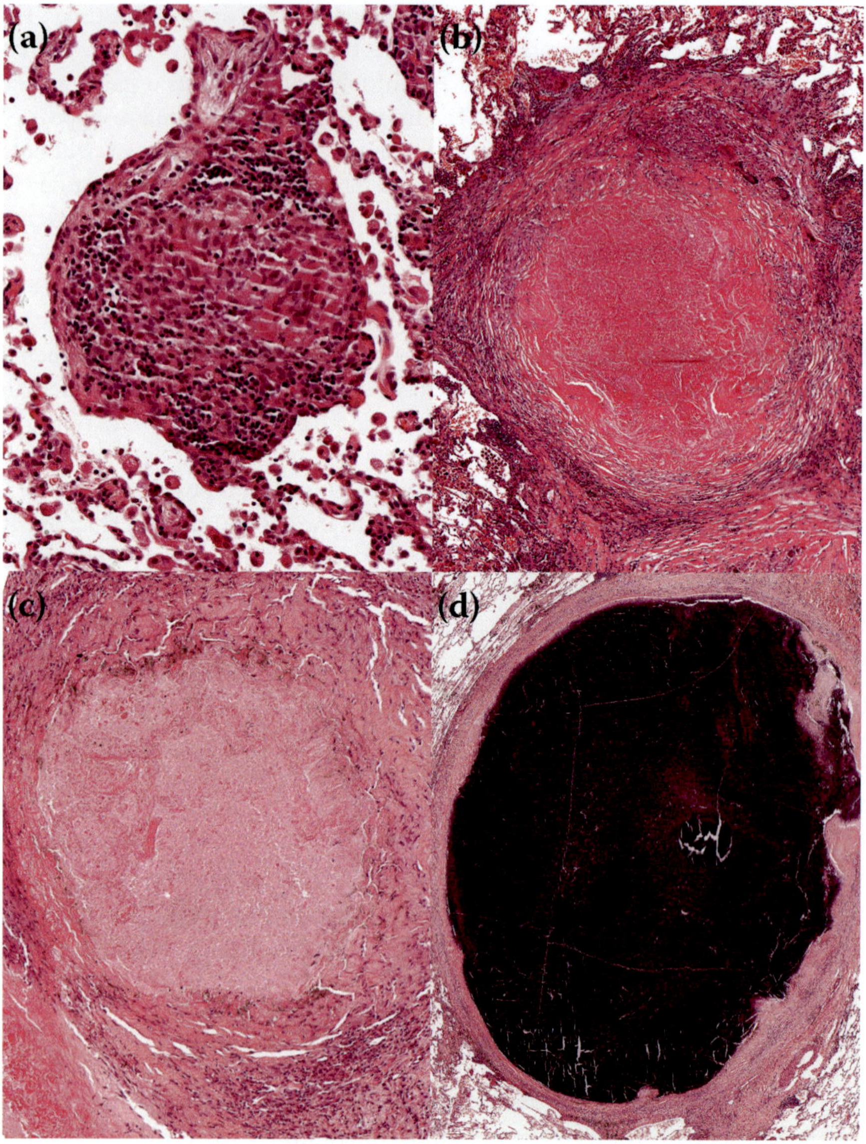

FIGURE 5.2 Different manifestation of tuberculous granulomas (H&E, LP). (a) Early nonnecrotizing granuloma. (b) Necrotizing granuloma. (c) Sclerotic granuloma. There is minimal inflammation, and the lesion is generally paucicellular (note the lack of basophilic staining). There is some activity in the lower right corner, possibly a mild chronic inflammatory cell infiltrate, but the granuloma itself lacks the multilayered appearance of an active necrotizing granuloma. (d) A calcified granuloma with sclerotic rim (no visible infectious activity). Note the almost complete absence of mononuclear infiltrates in the fibrotic rim.

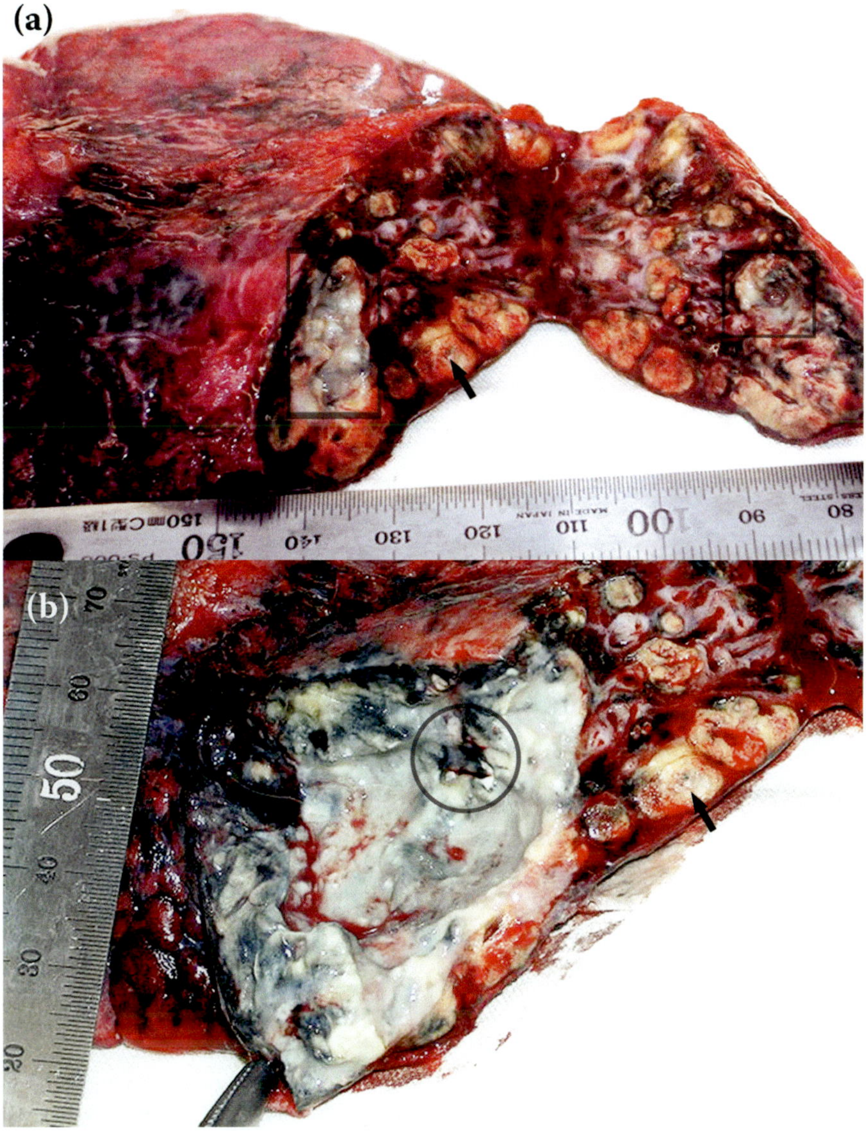

FIGURE 5.3 Upper lobectomy specimen demonstrating cavitating granuloma, multiple caseating granulomas, and secondary tuberculous pneumonia. (a) The lung has been sliced to reveal multiple granulomas (arrow) and a cavitating lesion (rectangle). Intervening lung tissue is consolidated and will show tuberculous pneumonia histologically. (b) Slicing into the cavity reveals it is more than 40 mm in maximum dimension. A possible communication with an airway is circled.

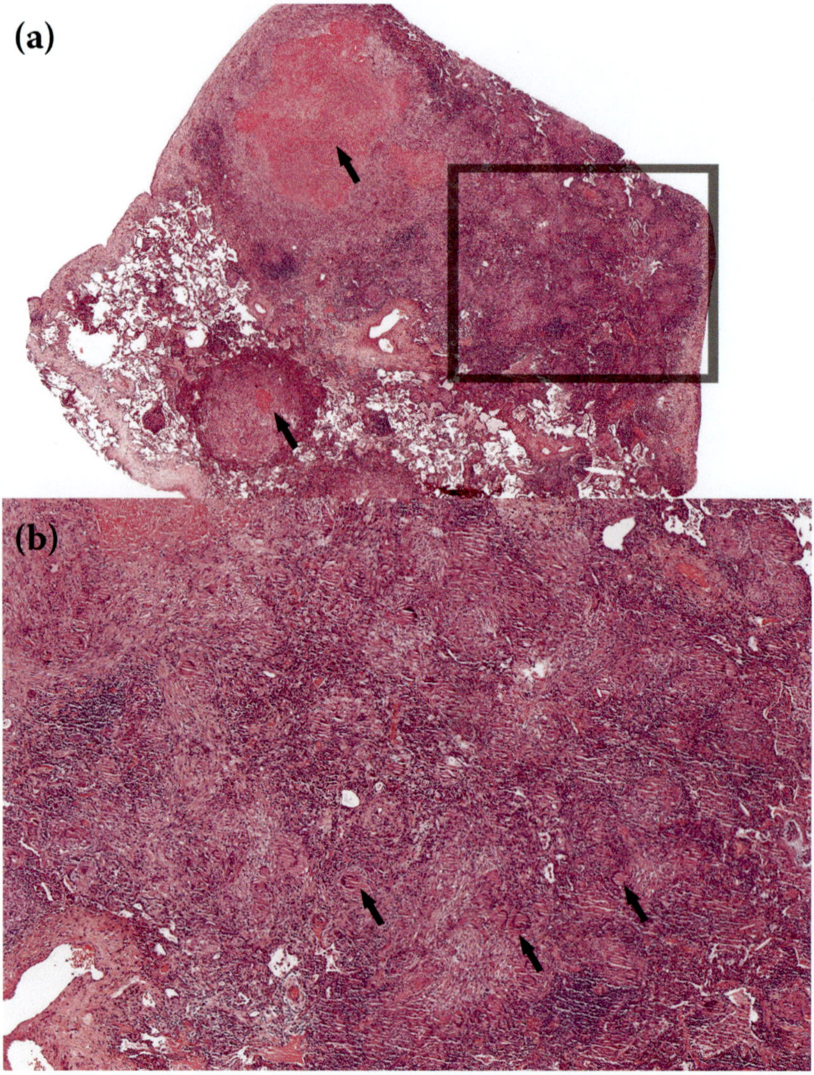

FIGURE 5.4 Tuberculous pneumonia (H&E, LP and MP). (a) A low-power view of a segment of lung demonstrates a large necrotizing granuloma (top arrow) with at least one large satellite lesion (lower arrow). The larger granuloma is poorly contained, and the adjacent lung is consolidated (rectangle). (b) A medium-power view of the rectangled area shows the consolidation to be composed of confluent aggregates of epithelioid histiocytes mixed with Langhans' and foreign-body type multinucleate giant cells (arrows). At this magnification it is possible to discern lymphoid aggregates (bluish areas). These are the same cells that make up a granuloma, but without a fibrotic wall to contain them they have spread to fill alveolar spaces and surrounded normal lung structures.

Central necrosis leads to cavitation of the lesion if there is communication with an adjacent bronchus or bronchiole. However, progressive fibrous encapsulation without cavitation is another possible outcome if there is no connection with an airway (Figure 5.6b). Severe fibrosis can result in retraction of the overlying pleural surface and can also appear blackened with anthracotic pigment (Jantz and Antony, 2006).

FEATURES

Macroscopic Features

Common surgical specimens include wedge excisions, subsegmental resections, lobectomies, and pneumonectomies. Surgical resection is usually undertaken for intractable MDR disease. In this clinical setting, it is most likely that the specimen resected will be inflamed and fibrotic with multiple secondary pathologies. If possible, fixation of the specimens should be performed with buffered formalin injected via the major bronchi (or into peripheral lung parenchyma if the bronchi are obstructed) with the option of additional insufflation of air at moderate pressure. After insufflation, the surgical specimens should be completely submersed in formalin and covered with a wet towel. Complete fixation of lobes or lungs takes at least 24 hours if the specimens are insufflated carefully. This may be shortened to 4 to 6 hours by a microwave-assisted technique (Kayser, 1992). Special precautions are required for TB lung tissue, given the risk of infection and the debatable effectiveness of formalin (Kappel et al., 1996), and in Masan, postsurgical dissection is performed under Biosafety Level 3 conditions (Richmond and McKinney, 1999; Centers for Disease Control and Prevention, 2010).

Bronchial involvement with TB can lead to stricture formation—distal bronchiectasis, atelectasis, and secondary infection are common subsequent complications. Peribronchial lymph nodes infiltrated with TB may infect bronchial mucous glands by direct extension. They can also penetrate a bronchial wall and erode into the lumen of the bronchus. If the patient has had longstanding disease and a history of antimicrobial treatment, it may be possible to identify one or more tuberculous cavities which may manifest as open cavities, or stellate scars representing past granulomas. There may also be metaplastic bone formation in the scar tissue (Kumar et al., 2010).

Tuberculoma is a term originally used to describe a tuberculous granuloma mass in the meninges. It is now also used to describe firm round

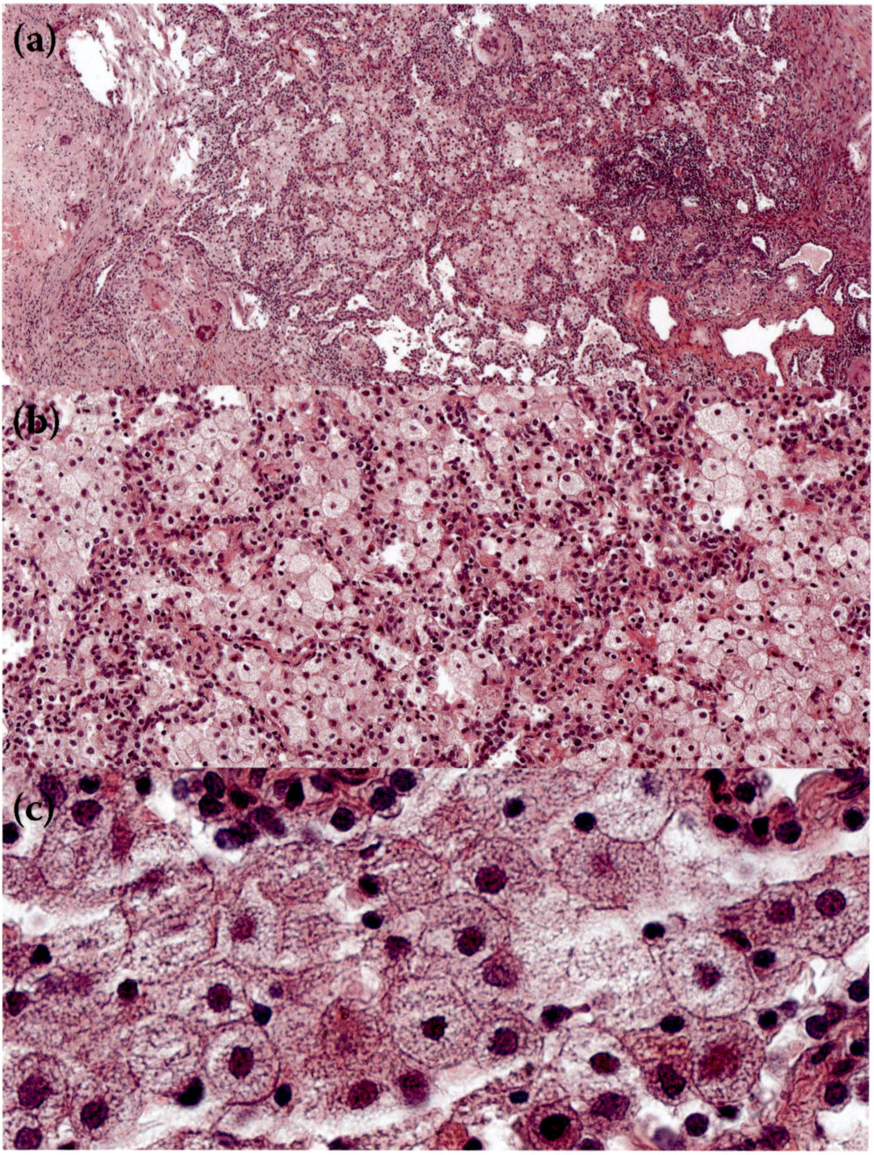

nodules, usually discrete or solitary in the lung, commonly located immediately below white or slightly yellowish pleura. Cutting through these may demonstrate concentric laminations, central calcification, or cavitation. The tuberculoma usually has a thick fibrous wall, and it is sometimes mistaken for a malignancy.

FIGURE 5.5 (opposite page) Foamy macrophages (H&E). (a, b, c) Foamy macrophages (foamy histiocytes, foam cells, xanthomatous cells) are macrophages which have phagocytosed cholesterol from damaged cells. Their cytoplasm consequently becomes very pale and bubbly in appearance. At low power (a), their cytoplasmic pallor is recognizable. At medium power (b), they can be seen in clusters with other inflammatory cells. At high power (c), their cytoplasm demonstrates an appearance similar to multiple trapped bubbles. Cytoplasmic margins are distinct. Nuclear-to-cytoplasmic ratio is low. Regarded mostly as a nonspecific finding which indicates tissue damage and repair, or chronic airway obstruction, they are not unique to TB, although there is speculation that they may play a more significant role in the development of secondary TB.

Microscopic Features

There are distinct differences between diagnostic histopathology and histopathology in research practice. In animal research studies, many of the parameters are controlled, and it is possible to draw firm conclusions because of this. In diagnostic histopathology, specimens are sometimes received without a confirmed diagnosis, often with minimal clinical history, and decisions are often made based solely on histopathological assessment. Consequently, descriptive terminology and reports have to be carefully worded. Consistent, clear, and precise phraseology is encouraged because this reduces the likelihood of misinterpretation. This is not as rigorously adhered to in research, although this should be encouraged.

There is no single microscopic feature which is truly pathognomonic of tuberculosis (El-Zammar and Katzenstein, 2007), although the observation of acid-fast bacilli in the histological setting of tuberculoid granulomas suggests a clear diagnosis of tuberculosis (Figure 5.7). Bacterial culture is confirmatory, but in view of the delay before such results are available, histological diagnosis is often sufficient for the initiation of treatment. In contrast, the diagnostic requirements for tuberculosis in the research setting are less stringent because the observed histopathology is often experimentally induced by the introduction of infecting mycobacteria (Stevens, 2002; Sugawara, 2009).

Acute (Early)

In primary TB, aerosolized bacilli are inhaled and tend to localize to the lung periphery. They elicit an initial transient neutrophil response. Many bacilli will survive this due to resistance conferred by their thick, glycolipid-rich bacterial cell wall. Some will be ingested by macrophages. Bacilli

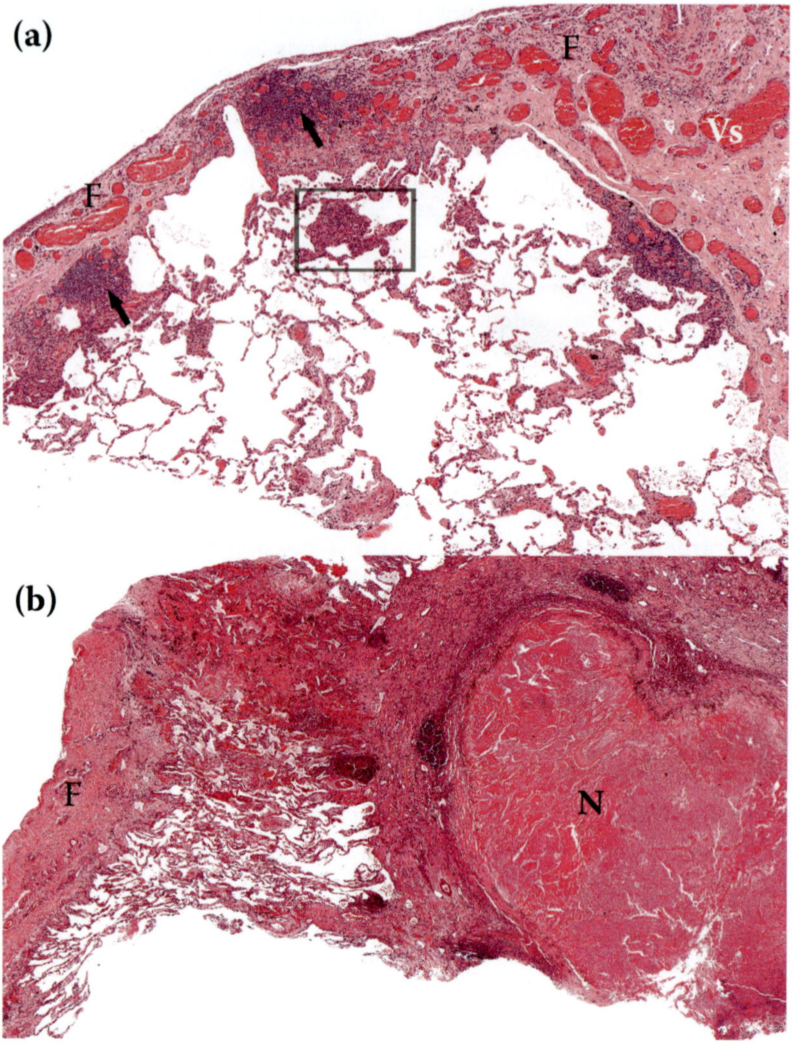

FIGURE 5.6 Pleural thickening (H&E, MP). Secondary pathological changes in the lung are often forgotten. (a) A low-power view of lung adjacent to a cavitating tuberculous granuloma (not shown) includes the pleural surface (top). There is significant fibrosis (F) extending along the pleural surface. Lymphoid collections (arrows) and numerous vessels (Vs) are within this. The adjacent lung is not entirely normal. A small nonnecrotizing granuloma (rectangle) is present. The degree of vascular congestion is most likely due to surgical artifact. (b) Pleural thickening (F) overlies a necrotizing granuloma (N). The fibrosis is mostly sclerotic collagen with scattered inflammatory cells. Pleural manifestation of TB is not rare, and is often misdiagnosed, because well-defined histiocytic granulomas are absent in the majority of cases.

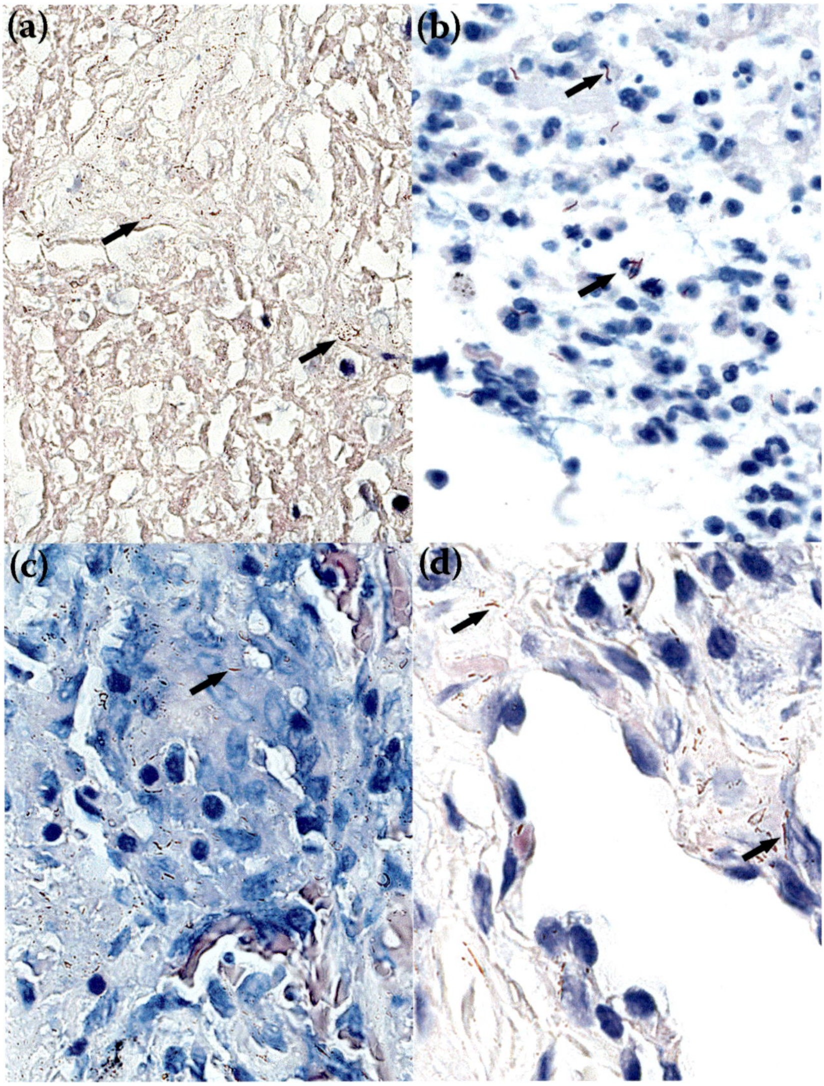

FIGURE 5.7 Acid-fast bacilli (ZN, HP oil). There is often discussion on the best place to look for acid-fast bacilli, but the truth is that they can be found anywhere. Here we demonstrate partially degenerate bacilli (acid-fast is magenta) within the necrotic debris inside a granuloma (a), at the interface between necrosis and normal lung (b, c), and also in adjacent lung tissue (d). Care should be taken when slicing unfixed lung tissue because the blade can carry bacilli, leading to erroneous conclusions regarding localization of bacteria. Examination under oil immersion objective is recommended. The demonstration of acid-fast bacilli in formalin-fixed material is often not possible, and polymerase chain reaction (PCR) examination of fresh tissue is recommended to confirm or exclude TB.

will also stimulate lymphocytes to become sensitized and produce various factors which in turn attract more macrophages and enhance their ability to kill ingested bacilli. These macrophages develop a granular eosinophilic cytoplasm and acquire a high cytoplasmic-to-nuclear ratio with a strong resemblance to epithelial cells. These form the basis of the early granuloma. Early tuberculous granulomas are small (several millimeters in diameter) and characterized by ball-like aggregates of epithelioid histiocytes with chronic inflammatory cells around the periphery (Figure 5.8). There may be central necrosis. At low power, a distinct but small cellular aggregate of lymphocytes can be observed in sharp contrast to the background lung parenchyma (Figure 5.9, Figure 5.10).

It is unusual to see an early lesion in isolation in human specimens, because by the time tissue is biopsied or resected, the granuloma has developed and progressed. However, it is possible to see early TB granulomas as satellite lesions in close proximity to much larger well developed granulomata. These arise due to lymphatic or vascular seeding of bacilli.

Chronic (Late, Developed)

As the lesion develops, some of the macrophages will fuse to produce large multinucleate giant cells called Langhans' multinucleate giant cells (Black and Epstein, 1974; Chambers, 1978). At the periphery, there will be a collection of lymphocytes, reflecting response of the immune system.

Central necrosis results in enlargement of the tubercle, and at the same time the zone of peripheral macrophages and lymphocytes becomes

FIGURE 5.8 (opposite page) Chronic inflammatory cells (H&E). Chronic inflammatory cells play a major role in manifestation and progression of TB. A nondisturbed interaction between the different cell types is a prerequisite for the formation of "classic epithelioid" granulomas, which are seldom present in TB infections of immunocompromised patients (for example, those suffering from hematologic malignancies). In conventional H&E stains, the different cell types include lymphocytes, macrophages, epithelioid histiocytes, eosinophils, and fibroblasts. (a) At low power, epithelioid histiocytes will appear pale due to their cytoplasm. In TB they will often cluster together (arrows). In contrast, lymphocytes are recognizable at low power as areas of dark blue. (b) At high power, epithelioid histiocytes are large cells (compare them with a red blood cell, rectangle) with pale abundant cytoplasm. Cytoplasmic margins are indistinct (arrow). They have nuclei which are vesicular and have prominent nucleoli. (c) Lymphocytes have dark blue nuclei and very little cytoplasm. (d) Plasma cells (rectangle) have eccentric nuclei with a clock-face appearance. Eosinophils (E) are less common and can be identified by their eosinophilic cytoplasmic granules. F—fibroblasts. Vs—vessel.

relatively thinner. Central "caseation" is a macroscopic description referring to the lumpy cottage cheese appearance visible to the naked eye. It should not be used for histological description. Histologically, the appropriate term is "necrotizing." Spindle-shaped fibroblasts appear in the peripheral lymphocyte zone with further development. These cells will lay down the extracellular collagenous tissue which becomes more prominent in older lesions (Figures 5.11 through 5.14).

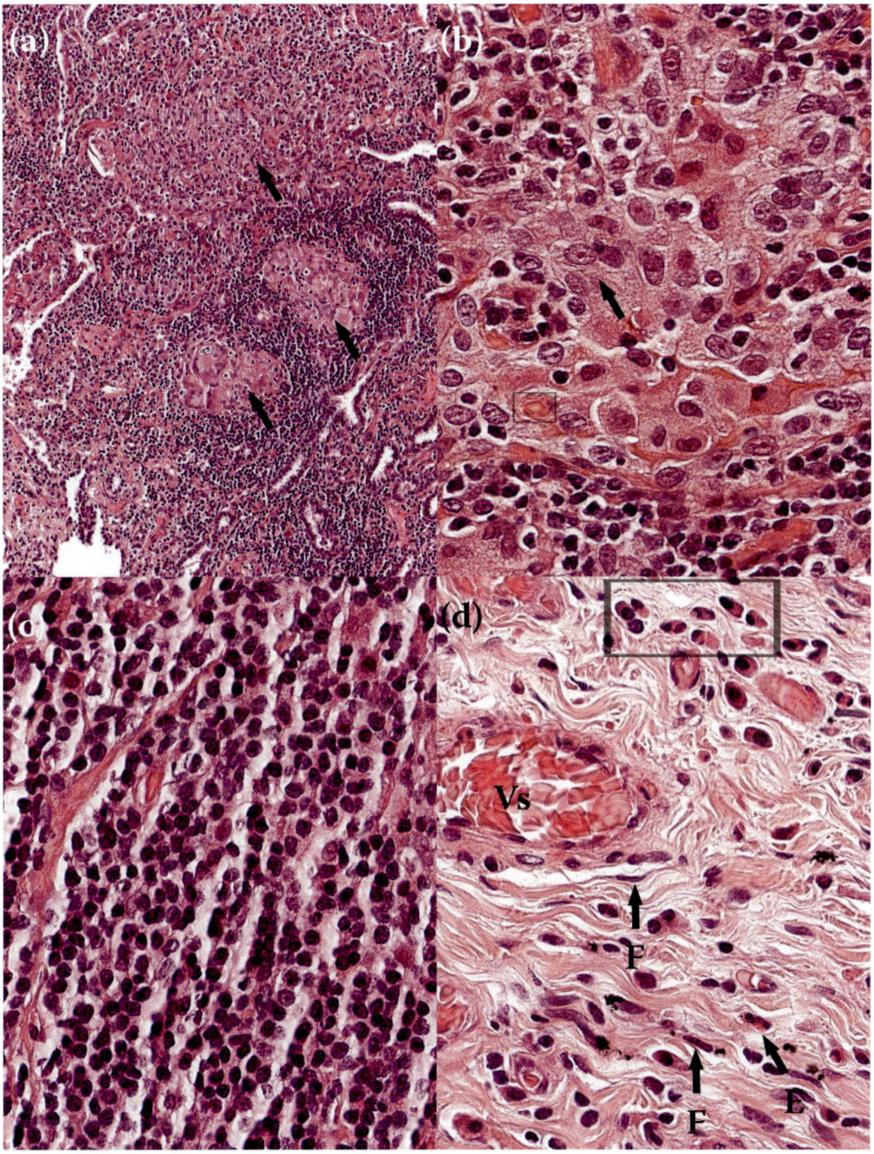

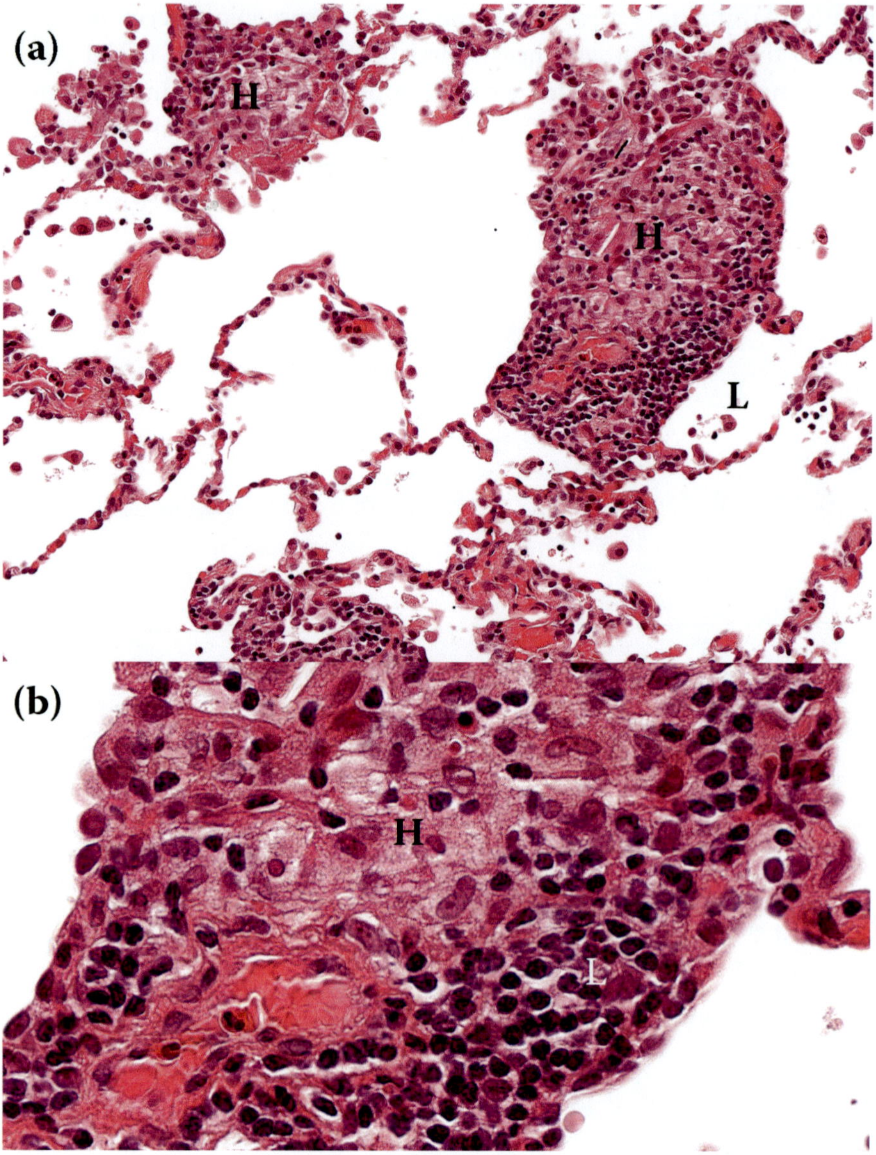

FIGURE 5.9 Nonnecrotizing granulomas (H&E, MP HP). (a) Medium-power view of two small nonnecrotizing granulomas which demonstrate ball-like aggregates of epithelioid histiocytes (H) with peripheral lymphoplasmacytic aggregate. (b) High-power view demonstrates epithelioid histiocytes with smooth granular nuclei and occasional nucleoli, large abundant cytoplasm, and ill-defined cytoplasmic margins. The periphery of the granuloma has a mixture of lymphocytes with occasional plasma cells (L). These appearances are not specific for tuberculosis and can be reproduced by a number of other infective and noninfective conditions.

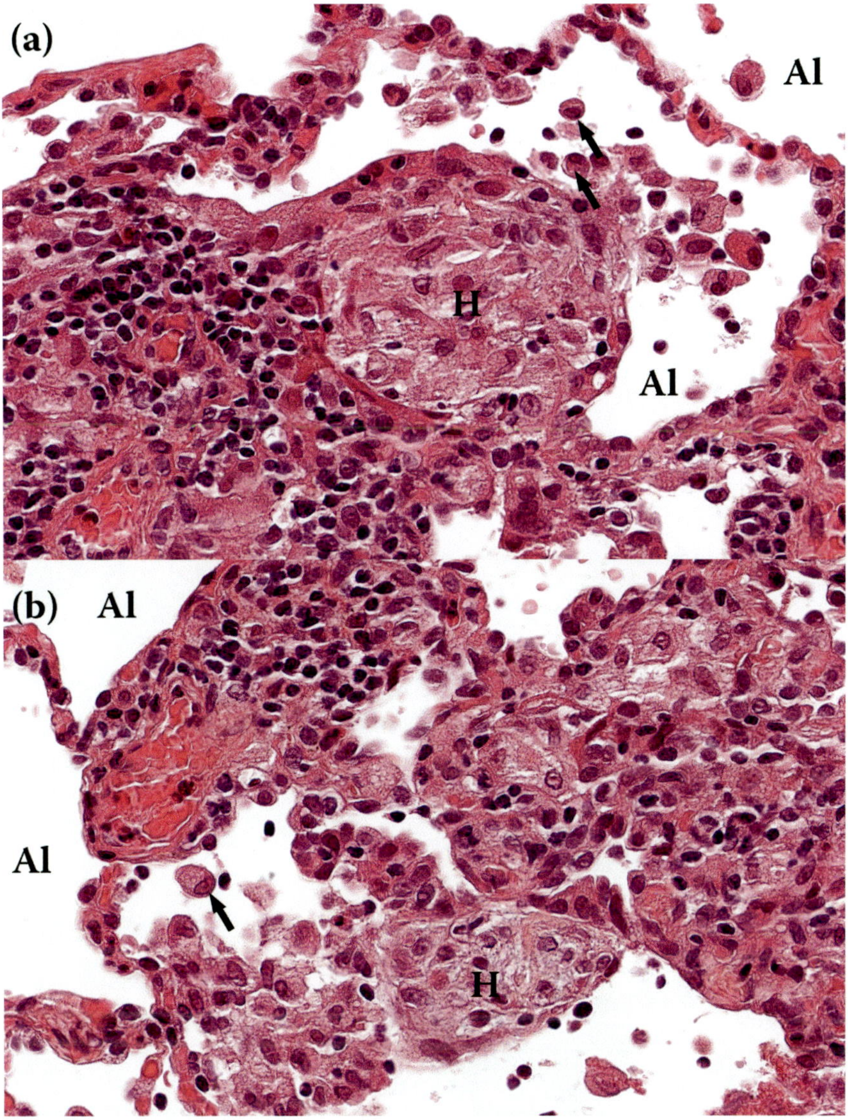

FIGURE 5.10 Early nonnecrotizing granulomas adjacent a developed tuberculous granuloma (H&E, HP). More examples of nonnecrotizing granulomas illustrating hematogenous seeding of infection to surrounding lung tissue. Although these granulomas do not demonstrate necrosis, it is incorrect to assume that only large lesions have necrosis. It is possible to find small (<1 mm diameter) necrotizing TB granulomas. Multinucleated giant cells are still absent; fibrous debris is present in the ill-defined granulomas. The close proximity of the granulomas (b) indicates an airborne infectious disease (in contrast to early sarcoidosis). Al—alveoli/alveolar duct. Arrows—alveolar histiocytes.

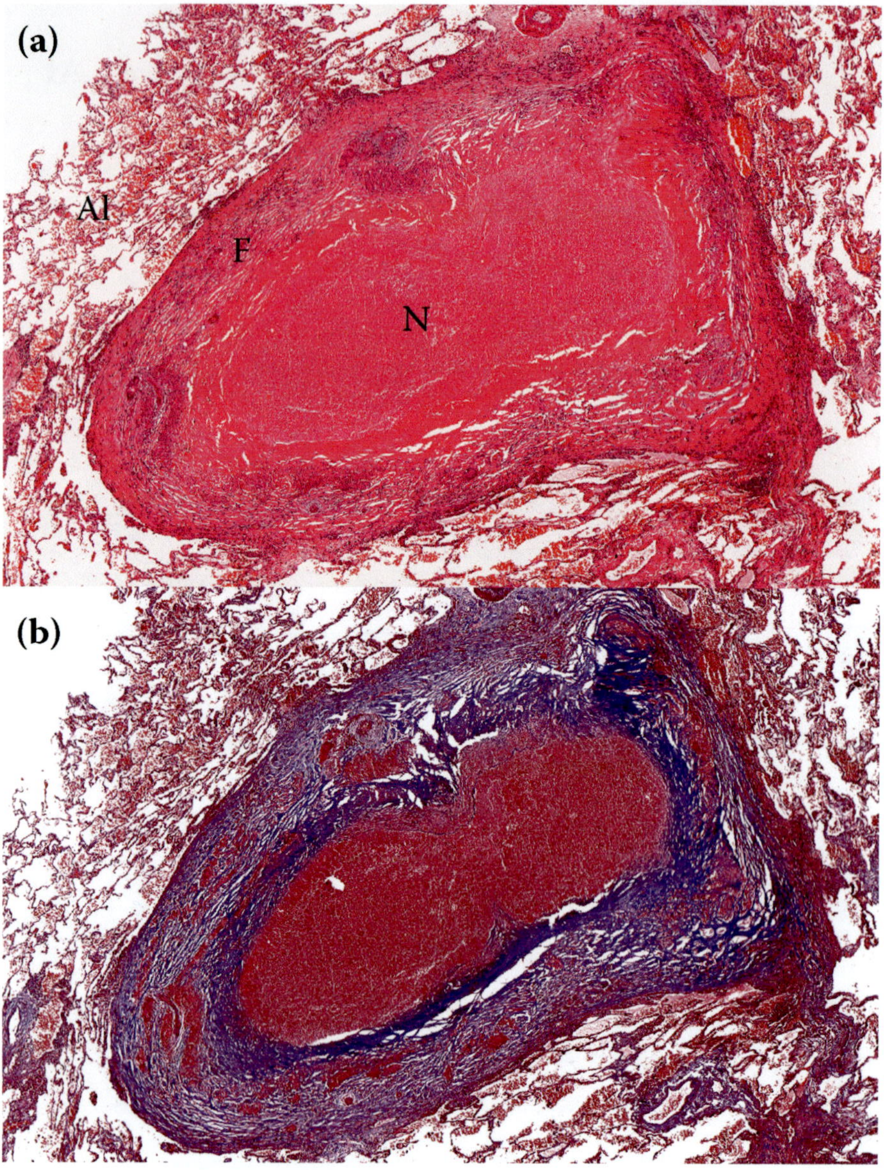

FIGURE 5.11 Necrotizing granuloma. (a) This low-power H&E-stained view of a necrotizing granuloma demonstrates central necrosis (N), a layer of cellular infiltrates (epithelioid histiocytes, chronic inflammatory cells, and multinucleated giant cells), as well as a fibrosis (F), which clearly defines the granuloma from surrounding lung alveoli (Al). (b) The Masson's trichrome stain of the same lesion shows the collagenous areas in blue.

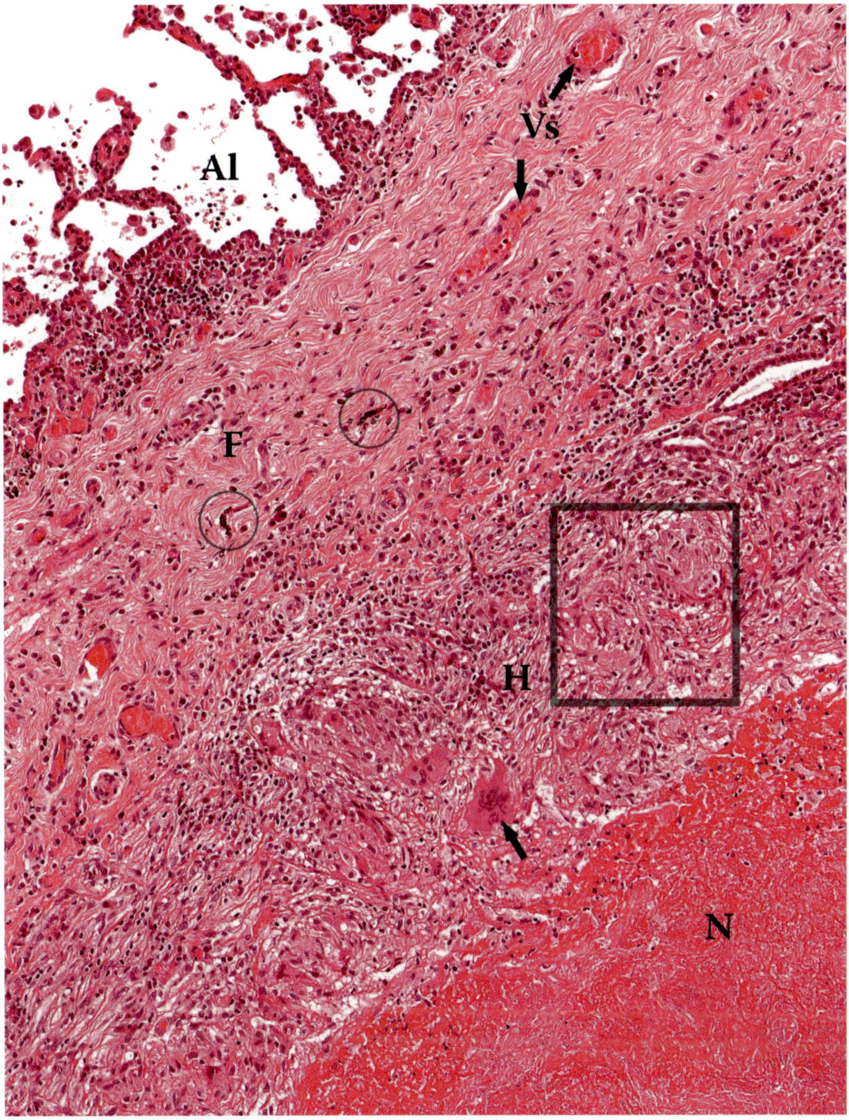

FIGURE 5.12 Layers of a granuloma (H&E, HP). Central necrosis appears as generally pink amorphous material, with cellular debris and ghost-like outlines of cells (N). Surrounding the necrosis is a layer of epithelioid histiocytes (H), which can be found in ball-like aggregates (rectangle) mixed with lymphocytes, plasma cells, and occasional eosinophils. There are scattered Langhans' and foreign-body type multinucleate giant cells (arrow). The outer layer is a fibrotic one (F) composed of fibroblasts, collagen, new vessels (Vs), and scattered chronic inflammatory cells. Focal anthracotic pigment (circles) can be found, often ingested by macrophages. Adjacent alveolar spaces (Al) are filled with alveolar macrophages.

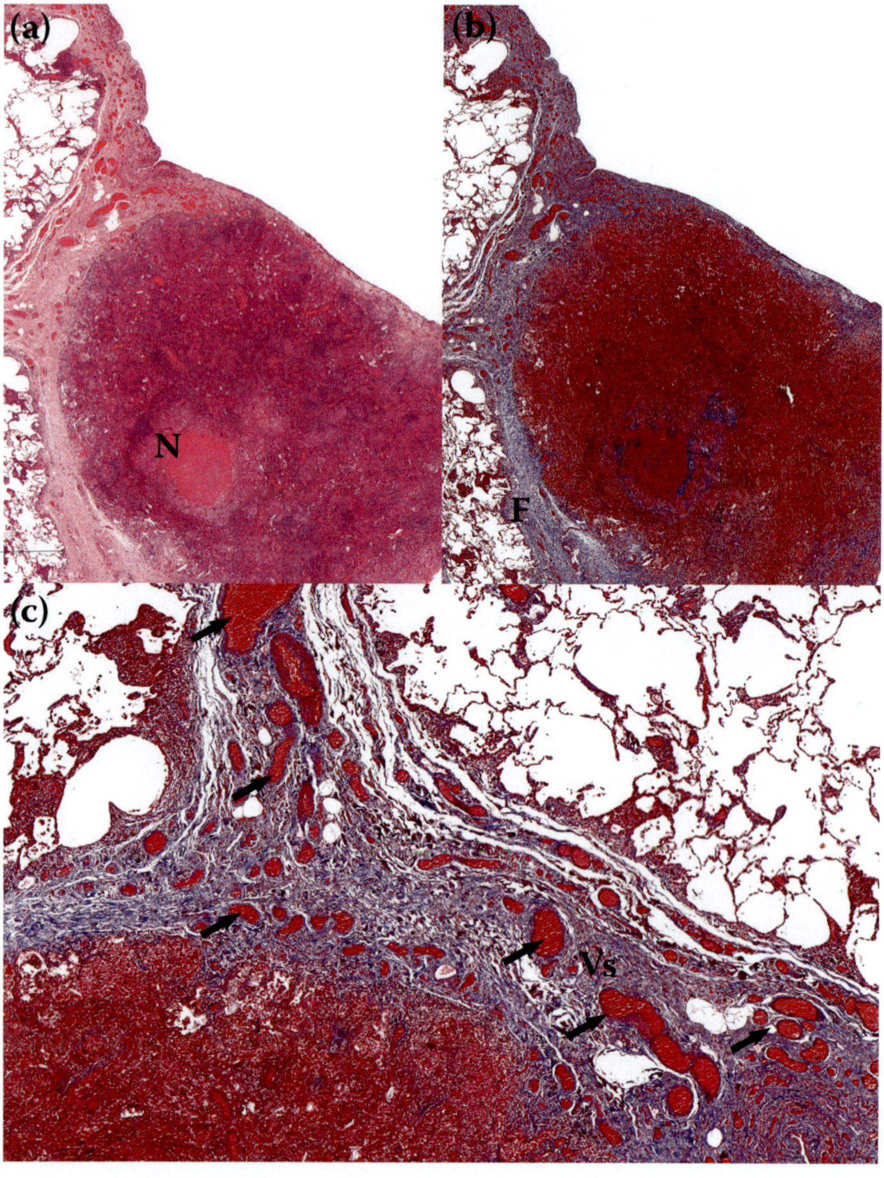

FIGURE 5.13 Fibrosis and vascularity around a TB granuloma (Masson's trichrome). (a, b) These two low-power images stained with H&E and Masson's trichrome demonstrate a large TB granuloma with a thick fibrous wall. (c) The fibrocollagenous wall (blue) which forms around a tuberculous granuloma may be millimeters in thickness and may display neovascularization (arrows).

Subsequent changes depend on the balance between the immune response and the infecting organisms (Flynn et al., 1993; Saunders and Cooper, 2000; Kaplan et al., 2003; Kaufmann, 2008). If the immune system is sufficiently strong, then the infection is contained within this macrophage–lymphocyte–fibroblast barrier, a barrier which is further strengthened by production of collagen by fibroblasts. If the infection is more aggressive or the immune system too weak, then the tubercle will rapidly enlarge due to increasing necrosis, and the infection is not contained. Tuberculous pneumonia is one possible outcome. Fully developed tuberculous granulomas tend to be rounded with pushing borders. At lower power multiple layers created by the different components of the granuloma (central necrosis, histiocytes, lymphocytes/plasma cells, occasional eosinophils, collagenous fibrosis) may be distinguished (Figures 5.11, 5.12, and 5.15). This is not diagnostic but highly characteristic. Sometimes irregular areas of confluent necrosis can occur in particularly large lesions or where lesions have coalesced (El-Zammar and Katzenstein, 2007).

At medium power it is possible to identify both necrotizing and non-necrotizing granulomas within the same section. Bronchiolar involvement is common. There are reports of airway-centered distribution of granulomas similar to bronchocentric granulomatosis (Takemura et al., 2000). In the human lung, it is common to see macrophages with anthracotic pigment (Colby and Yousem, 1997; Kumar et al., 2010).

At high power individual cells are clearly identifiable. Langhans' multinucleate giant cells are well-defined (Figure 5.16). Necrotic areas can contain amorphous granular debris or ghosts of alveolar septae ("infarct-like necrosis").

The presence of acid-fast bacilli does not make an unequivocal diagnosis of TB (as *Nocardia* and other species of *Mycobacterium* can also be acid-fast), but it does make it highly likely the lesion is due to TB. These acid-fast organisms are typically found in the necrotic debris and at the interface between necrotic and viable cells (Figure 5.7). Use of an oil-immersion microscope objective is required. An alternative to the Ziehl-Neelsen acid-fast stain is the auramine-rhodamine stain, which allows more sensitive fluorescence microscopy to be used for visualization.

Latency

Latency is a clinical term, not a histopathological one. It refers to a clinically asymptomatic individual who demonstrates immunological sensitization which has traditionally been defined by reactivity in the tuberculin skin test (TST) (Barry et al., 2009). Although latent granulomas have been

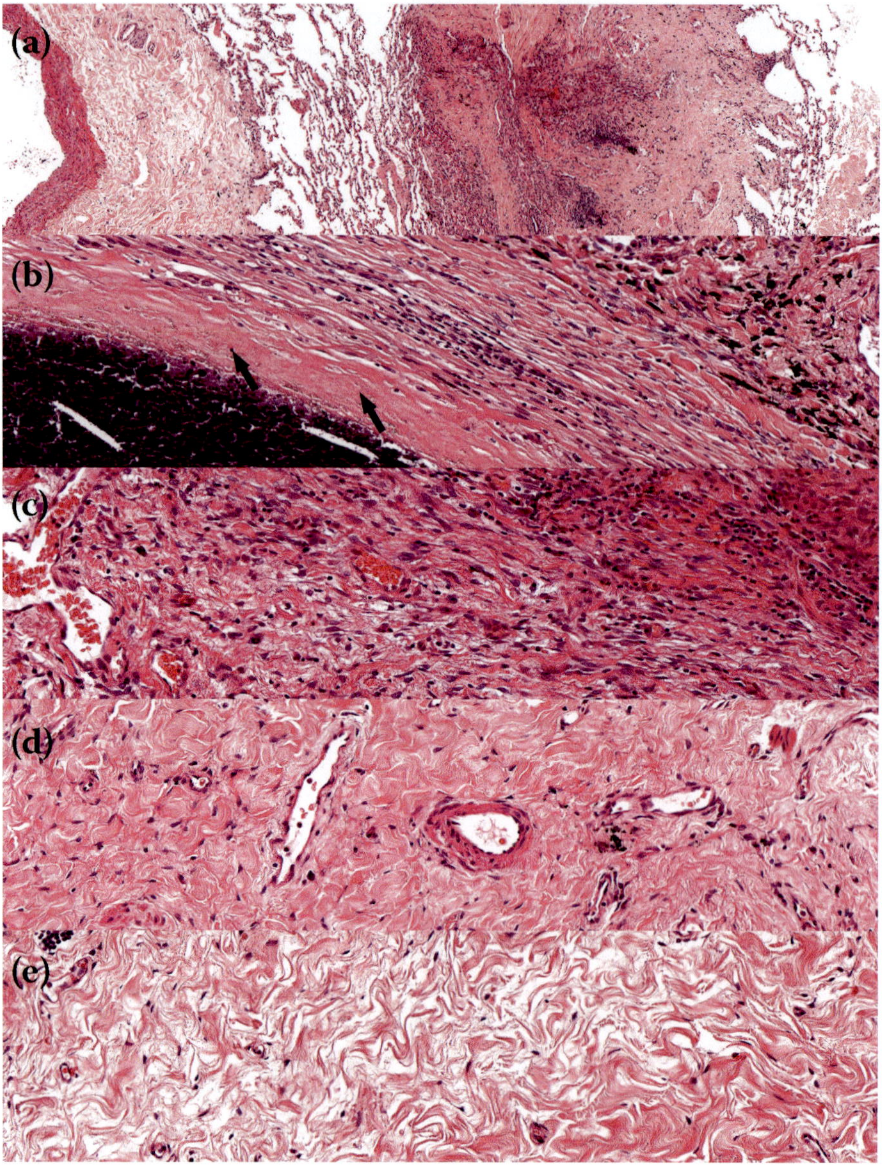

demonstrated in the experimental monkey model (Capuano et al., 2003; Flynn et al., 2003), histological appearances using histochemical stains alone give no indication as to the immunological status of the granuloma or whether there are viable bacilli still present. Such appearances do not correlate with any specific clinical testing, which is why use of the term "latent" in the context of histopathology is confusing.

FIGURE 5.14 (opposite page) Connective tissue and fibrosis, variations (H&E). Much is said about fibrosis in humans and the lack of it in some animal models, but it is not only the degree of fibrosis but the range of fibrotic changes that are remarkable. This panel illustrates how a novice may assess this with an H&E section by simply estimating the ratio of pink (cytoplasm, collagen, elastin, extracellular matrix), white (empty), and blue (nuclei or calcification). (a) LP. Pulmonary artery on the left and a scar on the right which, considering the clinical history, probably represents a healed granuloma. The scar is mostly pink, very little white, with a scattering of blue. There are no layers. This is not a granuloma but fibrocollagenous tissue which has contracted. Some chronic inflammatory cells (blue) are caught within this connective tissue. (b) MP. Wall of a calcified granuloma demonstrating sclerosis. The calcification is blue. Sclerosis is a change seen with aging which is predominantly collagen with very few fibroblasts (arrows). The lack of cells results in this appearing pink and solid. There are dense deposits of anthracotic pigment in the surrounding lung tissue (upper right), probably induced by heavy smoking. A definite diagnosis such as (inactive) TB cannot be derived from this lesion, which might also be caused by a fungal infection or healed infarction. (c) MP. Dense connective tissue. The ratio of pink:blue is approximately 3:1. This is connective tissue with a mixture of spindly fibroblasts and more active plump fibroblasts intermixed with chronic inflammatory cells (the blue dots in the upper right) and collagen. It is described as dense due to the lack of white space. Granulation tissue (the fibroinflammatory matrix seen in healing) would also have this appearance. Numerous fibroblasts have formed a nonspecific cellular fibrous lesion, which should not be mistaken for a granuloma. (d) MP. Dense connective tissue with minimal activity. A quiescent nonspecific fibrous tissue with some scattered mononuclear inflammatory cells, a frequent peribronchial and subpleural finding. In contrast to (c), the pink:blue ratio is higher and there is little white space. This connective tissue is mostly wavy elastin and collagen with scattered spindly fibroblasts. There are only a few chronic inflammatory cells. There are scattered small vessels. (e) MP. Loose connective tissue. The pink:white ratio is lower than in (d). The network of collagen and elastic and extracellular matrix is loosely packed, creating white spaces on tissue processing. There are very few inflammatory cells, and the fibroblasts are elongated and spindly. Several fragmented fibrous fibers can be seen, which sometimes are induced by (primary or secondary) amyloidosis.

Histologically the degree of inflammation, the cellularity, and presence of changes associated with healing or aging are assessed. While it is possible to describe a tuberculous granuloma as active based upon the presence of florid inflammation and/or associated lung features, it is not correct to assume that a granuloma with minimal inflammation

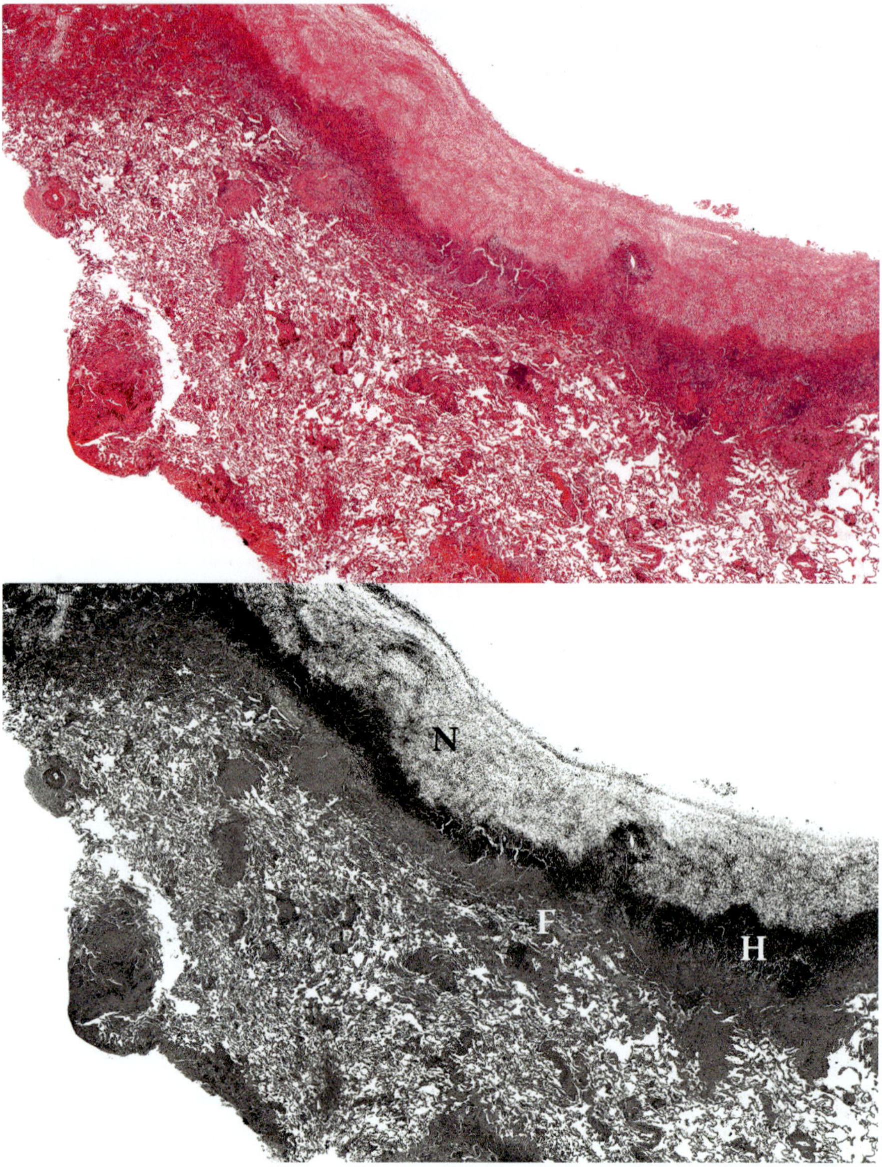

does not contain viable bacilli. Other terms such as "quiescent," "dormant," "inactive," or "paucicellular" may be encountered. "Dormant" and "inactive" are words which physicians regard as synonymous with "latent" and are best not employed when describing histology, where it is more appropriate to use only descriptive terminology, for example,

FIGURE 5.15 (opposite page) Wall of a cavitating tuberculous granuloma (H&E, LP). A cross section through the wall of a cavitating TB granuloma demonstrates a multilayered (trilaminar) appearance due to necrosis (N), histiocyte aggregates (H), and peripheral fibrosis (F). In other granulomas it may be possible to identify four layers because the necrotic core will demonstrate more friable material centrally, which often loosens during tissue processing. The layer of histiocytes and fibrosis can often be intermixed, and so at low power, perceiving a difference in texture rather than color is key. Separating these two layers is more easily seen in the lower image, which is a monochrome version in which textures have been accentuated. The fibrotic wall contains mononuclear infiltrates (including multinucleate giant cells) which indicates a still active and slowly progressing MTB lesion. Completely healed scars display a sharply demarcated fibrotic lesion without major cellular components.

"a granuloma with minimal inflammation and prominent sclerosis" (Figure 5.2c,d and Figure 5.5).

When a primary lesion heals, it may leave a scar with no evidence of the granuloma, or what remains may be a paucicellular structure with sclerosis and calcification. When a secondary lesion resolves following treatment the outcome may be similar. In both instances the histopathology is essentially the same—it is paucicellular, meaning most of the epithelioid histiocytes are gone and inflammatory cells are minimal or absent; there is fibrosis (as evidenced by the occurrence of fibroblasts, collagen, extracellular matrix) and/or sclerosis (as evidenced by the presence of collagen or extracellular matrix). Calcification may be seen. It is possible that the secondary lesion will be larger and/or multifocal, and the relative proportions of fibrosis, sclerosis, and calcification will vary. The size of the lesion and rate of healing are strong factors in determining outcome. Larger lesions tend to retain their rounded architecture with some shrinkage. Smaller lesions may disappear entirely. It is also possible to find granulomas of different appearances in the same lung, highlighting the heterogeneous nature of granulomas in the human.

A sclerotic granuloma is one in which most of the lesion is composed of paucicellular concentric sclerosis with a minimal chronic inflammatory cell infiltrate. A calcified granuloma is one in which mineralization may have replaced all but the periphery of the lesion (Figure 5.2). Sometimes only scar tissue may represent the site of a previous granuloma and occasionally focal calcification will be found in an active granuloma.

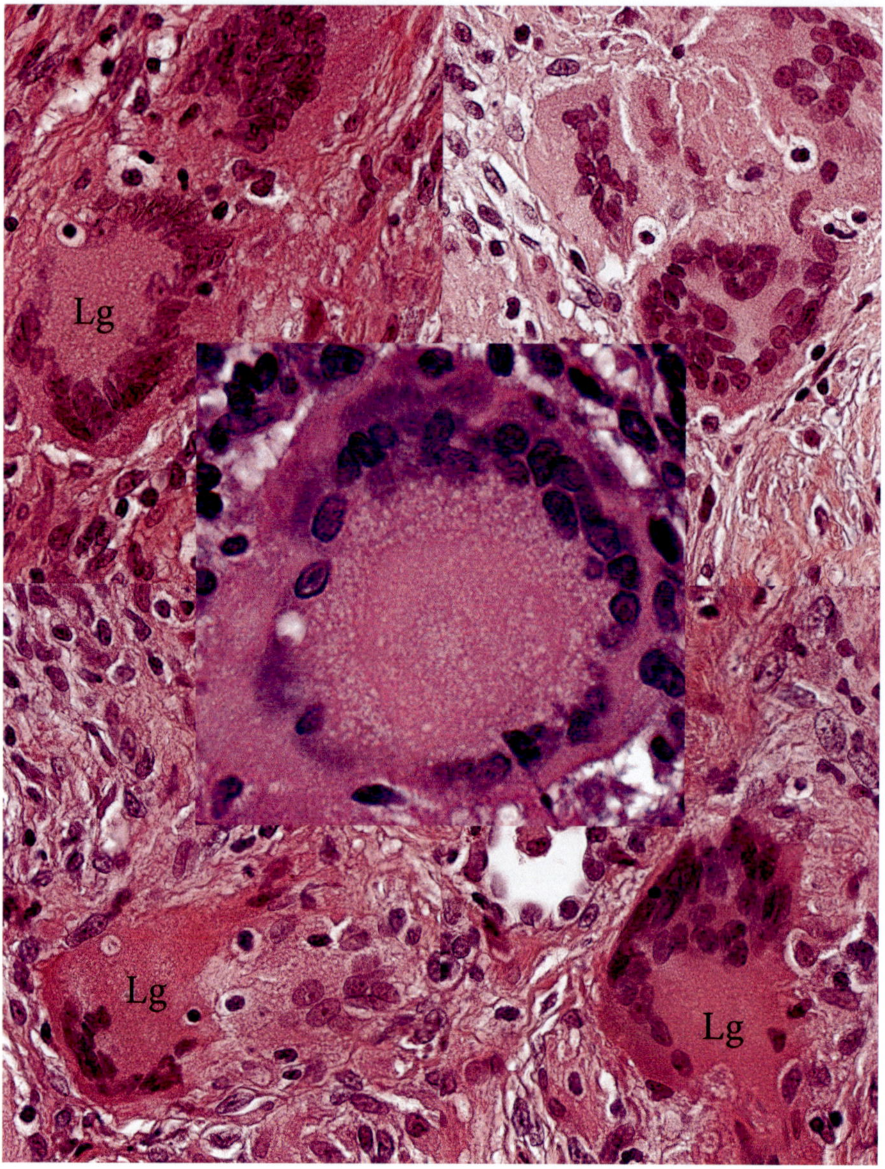

VARIATIONS

Human tuberculous granulomas are heterogeneous in appearance and it is common to find more than one subtype of lesion within the same lung segment. In malignancy histological variants can be associated with different outcomes. This has yet to be demonstrated in TB, so at present such subclassification is only undertaken for academic purposes. In addition to

FIGURE 5.16 (opposite page) Giant cell variations. This composite figure demonstrates a classic Langhans' type multinucleate giant cell centrally. Such cells have multiple nuclei arranged in a horseshoe shape at the periphery and are a common finding in TB granulomas. Surrounding this image are several other multinucleate giant cells from TB granulomas, of which three (Lg) meet the criteria for Langhans' type multinucleate giant cells. The remainder are best described as foreign-body type giant cells. Multinucleate giant cells are created from multiple histiocytes/macrophages and are a response to indigestible material. The significance of Langhans' type giant cells has been called into question, because they are not specific for TB and can be seen in a broad variety of chronic infectious or inflammatory lung diseases such as aspergillosis, sarcoidosis, Wegener's granulomatosis, or damage due to inhaled substances (berylliosis). Angle of sectioning through the giant cell has influence on its histologic appearance.

the varieties of sclerosed fibrocalcific TB granulomas already described, it is possible to find a nonnecrotizing solid granuloma, which is a granuloma without necrosis dominated by epithelioid histiocytes. This can be seen early in granuloma development, but such lesions may also grow to significant size. A suppurative granuloma is one in which neutrophils aggregate in the center of the lesion.

Complications of uncontrolled TB infection include tuberculous pneumonia (Figure 5.4)—which occurs when TB infection spreads throughout large areas of lung parenchyma as a diffuse bronchopneumonia. This is characterized by the diffuse spread of epithelioid histiocytes, multinucleate giant cells, and other inflammatory cells extending beyond the bounds of the granuloma and infiltrating peribronchiolar and alveolar spaces. Areas of necrosis can be large, and in severe cases, distinct granulomas are not easily recognized. Miliary tuberculosis refers to multifocal, often multiorgan infection which has spread via blood vessels and lymphatic channels. On gross examination, individual lesions are usually small (1 to 5 mm maximum dimension), distinct, yellow-white firm areas scattered like millet seeds throughout an organ. At a microscopic level, numerous necrotizing confluent granulomas can be found.

There are certainly more variations and combinations as these are descriptive terms that have not been standardized. The human inflammatory reaction is not monomorphic—it has a range of dynamic histological manifestations. The diversity and heterogeneity of lesions can be attributed to the age of the lesion, variability in the immunocompetency of the particular human host, and also the phenotypic characteristics of the TB bacillus (Russell, 2007).

REFERENCES

Andreu, J., Mauleon, S., Pallisa, E., Majo, J., Martinez-Rodriguez, M., and Caceres, J. 2002. Miliary lung disease revisited. *Curr Probl Diagn Radiol,* 31, 189–197.

Barry, C.E., 3rd, Boshoff, H.I., Dartois, V., Dick, T., Ehrt, S., Flynn, J., Schnappinger, D., Wilkinson, R.J., and Young, D. 2009. The spectrum of latent tuberculosis: Rethinking the biology and intervention strategies. *Nat Rev Microbiol,* 7, 845–855.

Black, M.M., and Epstein, W.L. 1974. Formation of multinucleate giant cells in organized epitheloid cell granulomas. *Am J Pathol,* 74, 263–274.

Capuano, S.V., 3rd, Croix, D.A., Pawar, S., Zinovik, A., Myers, A., Lin, P.L., Bissel, S., Fuhrman, C., Klein, E., and Flynn, J.L. 2003. Experimental *Mycobacterium tuberculosis* infection of cynomolgus macaques closely resembles the various manifestations of human M. tuberculosis infection. *Infect Immun,* 71, 5831–5844.

Centers for Disease Control and Prevention. 2010. *Laboratory biosafety level criteria* [online]. Centers for Disease Control. Available from: http://www.cdc.gov/OD/ohs/biosfty/bmbl4/bmbl4s3.htm [Accessed January 7, 2010].

Chambers, T.J. 1978. Multinucleate giant cells. *J Pathol,* 126, 125–148.

Colby, T.V., and Yousem, S.A. 1997. Lungs. In *Histology for pathologists,* ed, S.S. Sternberg. New York: Lippincott-Raven, 433–458.

El-Zammar, O.A., and Katzenstein, A.L. 2007. Pathological diagnosis of granulomatous lung disease: A review. *Histopathology,* 50, 289–310.

Flynn, J.L., Capuano, S.V., Croix, D., Pawar, S., Myers, A., Zinovik, A., and Klein, E. 2003. Non-human primates: A model for tuberculosis research. *Tuberculosis (Edinb),* 83, 116–118.

Flynn, J.L., Chan, J., Triebold, K.J., Dalton, D.K., Stewart, T.A., and Bloom, B.R. 1993. An essential role for interferon gamma in resistance to *Mycobacterium tuberculosis* infection. *J Exp Med,* 178, 2249–2254.

Ghon, A. 1912. *Der primäre Lungenherd bei der Tuberkulose der Kinder (The primary lung lesion in infant TB)* Berlin and Vienna: Urbach and Scharzenburg.

Hunter, R.L., Jagannath, C., and Actor, J.K. 2007. Pathology of postprimary tuberculosis in humans and mice: Contradiction of long-held beliefs. *Tuberculosis (Edinb),* 87, 267–278.

Jantz, M.A., and Antony, V.B. 2006. Pleural fibrosis. *Clin Chest Med,* 27, 181–191.

Kaplan, G., Post, F.A., Moreira, A.L., Wainwright, H., Kreiswirth, B.N., Tanverdi, M., Mathema, B., Ramaswamy, S.V., Walther, G., Steyn, L.M., Barry, C.E., 3rd, and Bekker, L.G. 2003. *Mycobacterium tuberculosis* growth at the cavity surface: A microenvironment with failed immunity. *Infect Immun,* 71, 7099–7108.

Kappel, T.J., Reinartz, J.J., Schmid, J.L., Holter, J.J., and Azar, M.M. 1996. The viability of *Mycobacterium tuberculosis* in formalin-fixed pulmonary autopsy tissue: Review of the literature and brief report. *Hum Pathol,* 27, 1361–1364.

Kaufmann, S.H.E., ed. (2008) *Handbook of tuberculosis: Immunology and cell biology.* Weinheim: WILEY-VCH Verlag GmbH & Co. KGaA.

Kayser, K. 1992. *Analytical lung pathology.* Berlin, New York: Springer-Verlag.

Kumar, V., Abbas, A.K., Fausto, N., and Aster, J. 2010. *Robbins and Cotran pathologic basis of disease, professional edition,* 8th ed. Philadelphia: Saunders/Elsevier.

Park, S.K., Kim, J.H., Kang, H., Cho, J.S., and Smego, R.A., Jr. 2009. Pulmonary resection combined with isoniazid- and rifampin-based drug therapy for patients with multidrug-resistant and extensively drug-resistant tuberculosis. *Int J Infect Dis,* 13, 170–175.

Park, S.K., Lee, C.M., Heu, J.P., and Song, S.D. 2002. A retrospective study for the outcome of pulmonary resection in 49 patients with multidrug-resistant tuberculosis. *Int J Tuberc Lung Dis,* 6, 143–149.

Richmond, J.Y., and McKinney, R.W., eds. 1999. *Biosafety in microbiological and biomedical laboratories.* Washington, DC: U.S. Government Printing Office.

Russell, D.G. 2007. Who puts the tubercle in tuberculosis? *Nat Rev Microbiol,* 5, 39–47.

Russell, D.G., Cardona, P.J., Kim, M.J., Allain, S., and Altare, F. 2009. Foamy macrophages and the progression of the human tuberculosis granuloma. *Nat Immunol,* 10, 943–948.

Saunders, B.M., and Cooper, A.M. 2000. Restraining mycobacteria: Role of granulomas in mycobacterial infections. *Immunol Cell Biol,* 78, 334–341.

Stevens, A., ed. 2002. *Wheater's basic histopathology: a colour atlas and text.* Edinburgh: Churchill Livingstone.

Sugawara, I. 2009. Why does tuberculosis lead to specific inflammation? *Nihon Hansenbyo Gakkai Zasshi,* 78, 263–269.

Takemura, T., Akiyama, O., Yanagawa, T., Ikushima, S., Ando, T., and Oritsu, M. 2000. Pulmonary tuberculosis with unusual cystic change in an immunocompromised host. *Pathol Int,* 50, 672–677.

CHAPTER 6

Pulmonary Tuberculosis in Monkeys

JoAnne L. Flynn and Edwin Klein

CONTENTS

INTRODUCTION

Nonhuman primates (NHPs), generally macaques, have been used for decades as a model of tuberculosis, for drug and vaccine studies (Good, 1968). There is a long history of "natural" *Mycobacterium tuberculosis* (MTB) or *M. bovis* infections in captive nonhuman primates, although rigorous screening practices have brought this under control. The nonhuman primate is genetically close to humans, and most human immunologic reagents are cross-reactive with macaques, which facilitates the use of this model for studying tuberculosis. Our studies (Capuano et al., 2003; Flynn et al., 2005; Lin et al., 2009), and those of others (Langermans et al., 2001, 2005; Chen et al., 2009; Verreck et al., 2009), have demonstrated that macaques recapitulate all manifestations of human MTB infection. This includes presentation of both latent and active infection, as well as reactivation tuberculosis. The pathology and granuloma structures observed are remarkably similar to those in human tuberculosis (Figure 6.2).

INOCULUM AND ROUTE OF INFECTION

A respiratory route of infection is used in macaques, either intrabronchial instillation or aerosol delivery. Our studies have used intrabronchial administration in cynomolgus macaques (*Macaca fasicularis*). Others have used rhesus macaques (*M. mulatta*); these animals may be more susceptible to active tuberculosis. Inoculum varies greatly among studies. For vaccine studies, most groups have used 1000 to 3000 CFU, a very high dose (Langermans et al., 2001, 2005). Our studies are generally with a low dose, 25 CFU, which leads to equal proportions of monkeys developing active disease or latent infection (Capuano et al., 2003; Lin et al., 2009). The incidence of active tuberculosis increases with higher inocula. Most macaque studies have used MTB strain Erdman, although H37Rv is also used. In the low-dose model, active TB usually develops within 6 months of infection; we categorize monkeys as having active disease or latent infection 6 to 8 months after infection. With higher inocula, monkeys can progress to active disease within 10 weeks.

FEATURES

Acute (Early)

The early critical period that defines infection outcome in this model appears to occur within the first 8 weeks post-inoculation, with microscopic disease sometimes becoming apparent in as little as 4 to 5 weeks (Lin, 2006). The very earliest morphological manifestation may be an amorphous aggregation of epithelioid histiocytes, although this has only been noted early in the draining lymph node. These likely progress to foci of granulomas—nodular inflammatory infiltrates with clear structural organization. The cellular composition of the granulomas directly reflects the (variable) underlying immunological containment processes

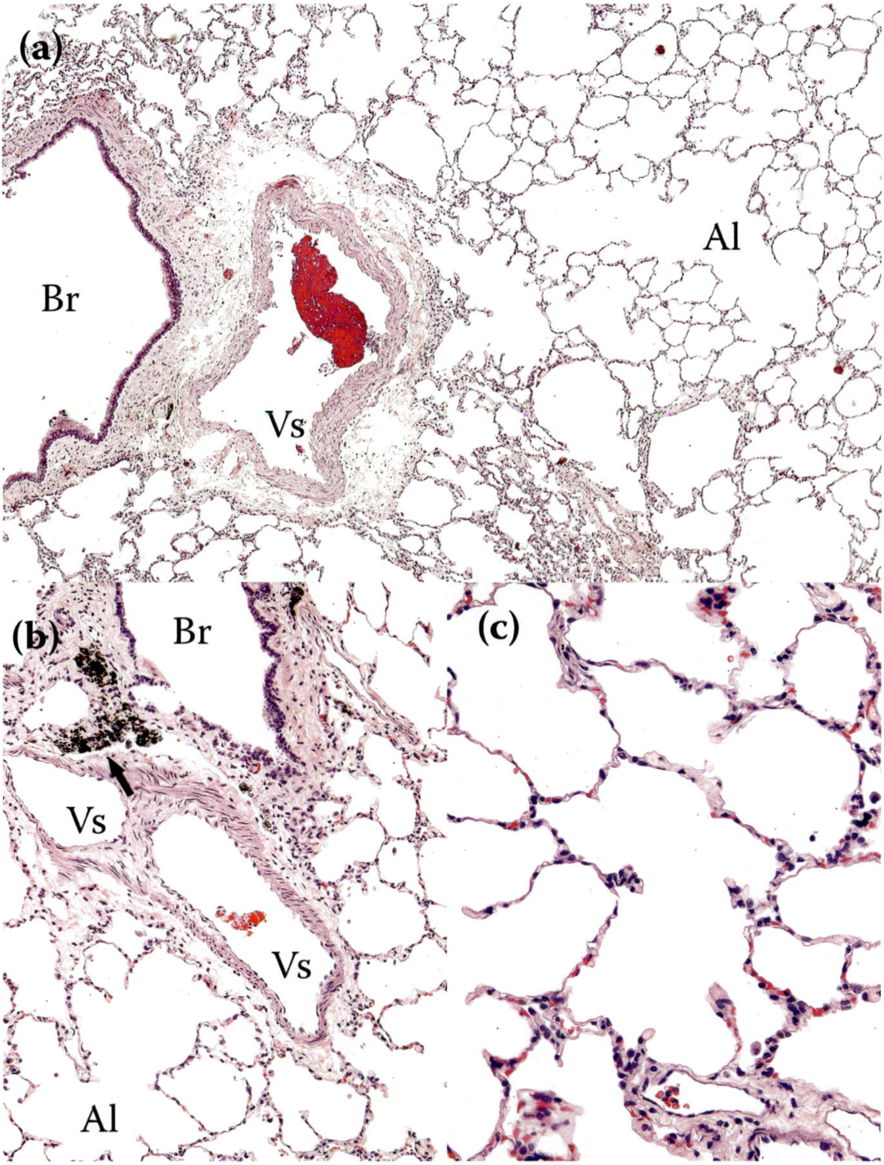

FIGURE 6.1 Normal lung (H&E). Normal arterioles (Vs) and alveolar spaces (Al) are shown at low- (a), medium- (b), and high-power (c) magnifications. In (b), patchy carbon pigment (arrow) is present adjacent to a bronchus (Br).

occurring. Granulomas may initially develop in a nonnecrotizing form, with a solid core of plump, epithelioid histiocytes and multinucleated giant cells without evidence of central necrosis (Figure 6.2b and Figure 6.3). Such lesions, both in lung and lymph node, are almost always associated with early, rapidly disseminating disease. In most cases however, initial lesions are necrotizing (Figure 6.2a, Figure 6.4, and Figure 6.5), with central eosinophilic amorphous debris corresponding to the caseous material seen grossly. Surrounding the necrosis is a rim of large, plump epithelioid histiocytes. These cells may syncytialize, forming typical Langhans' type giant cells with archiform pattern nuclei (Figure 6.5). Peripheral to this zone is a mantle of lymphocytes. Some granulomas demonstrate abundant numbers of admixed eosinophils along outer margins (Figure 6.5), although their role and the pathogenesis of recruitment are not apparent. At the outermost circumference, a margin of surrounding fibrosis may develop, depending on the stage and extent of containment of the lesion. If immunological containment is not realized subsequent to initial infection, disease spread can occur through a variety of mechanisms having different histomorphological manifestations. Viable organisms may escape the localized inflammatory nodule, extending directly into adjacent alveolar structures. This is seen microscopically as local alveolar extension, where histiocytes are noted infiltrating directly from the body of a necrotizing granuloma into adjoining alveoli (Figure 6.6). Another manifestation of early, fulminant disease dissemination involves a lesion referred to as the alveolar histiocytic focus. These foci, often associated with areas of granular fibrinoid necrotic debris are seen in terminal air spaces without associated adjacent granulomas (Figure 6.7). Such distribution, in conjunction with the fact that these are commonly noted in monkeys with rapidly evolving disease and/or very high quantitative bacterial loads, suggests that they may be one of the earliest lesions to manifest in association with hematogenous dissemination.

Chronic

Lesions evolving in a longer-term fashion will vary, depending on the balance of immunological containment achieved. In animals not succumbing to rapidly progressive disease within the first 3 to 5 months, granuloma growth and infection spread may evolve at a steady but progressive low-grade rate for over a year, prior to clinical deterioration. Disease present typically reflects both active necrosis and partial healing and containment. The latter is typically characterized by a degree of

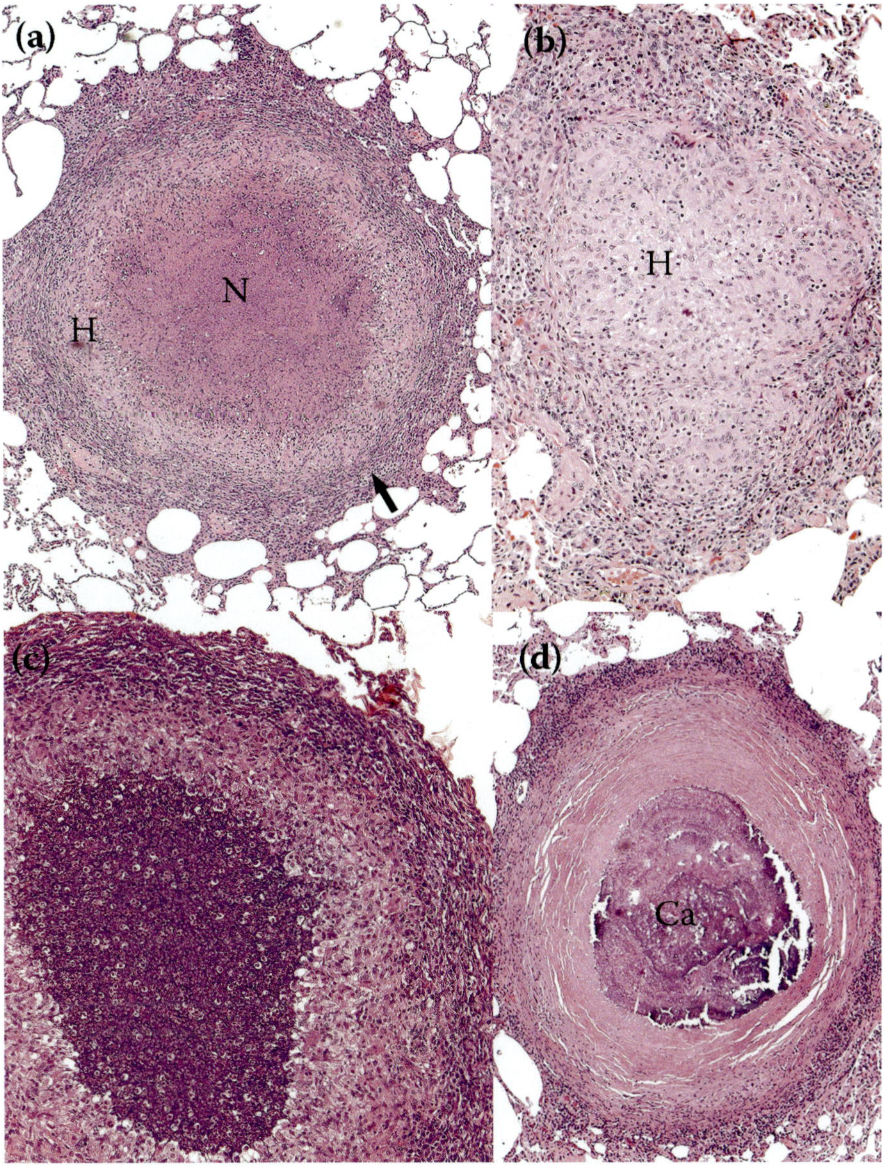

FIGURE 6.2 Granuloma variations (H&E, MP). (a) Typical necrotizing granuloma—note central eosinophilic necrotic debris (N) surrounded by a rim of large epithelioid histiocytes (H) and a thin zone of lymphoplasmacytic cells (arrow). (b) Typical nonnecrotizing granuloma comprised of a solid core of histiocytic cells (H). (c) Typical suppurative granuloma—note the central core of granulocytic cells. (d) Typical fibrocalcific granuloma—an extensive central area of calcification (mineralization, Ca) is surrounded by dense peripheral fibrosis.

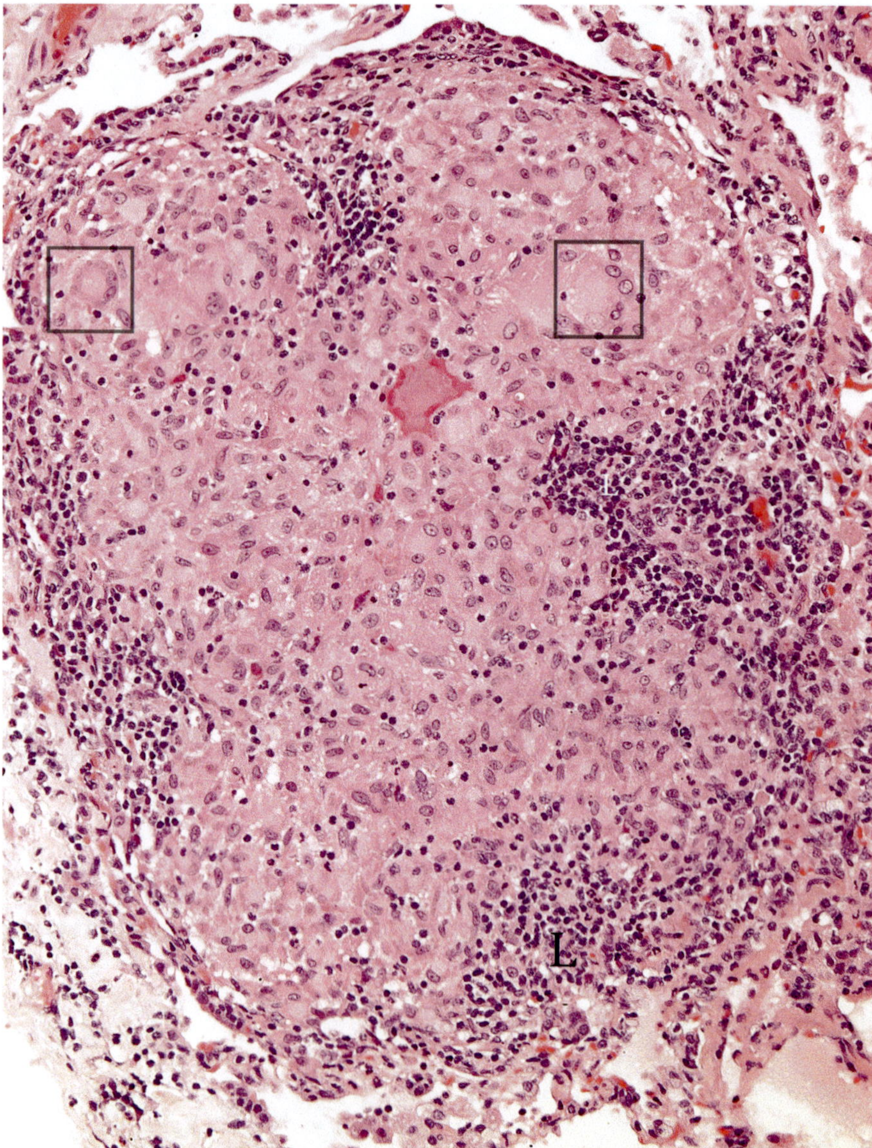

FIGURE 6.3 Nonnecrotizing granuloma (H&E, HP). Large epithelioid histiocytes occasionally syncytialized into multinucleated giant cells (squares), with patchy mild lymphoplasmacytic infiltrates (L).

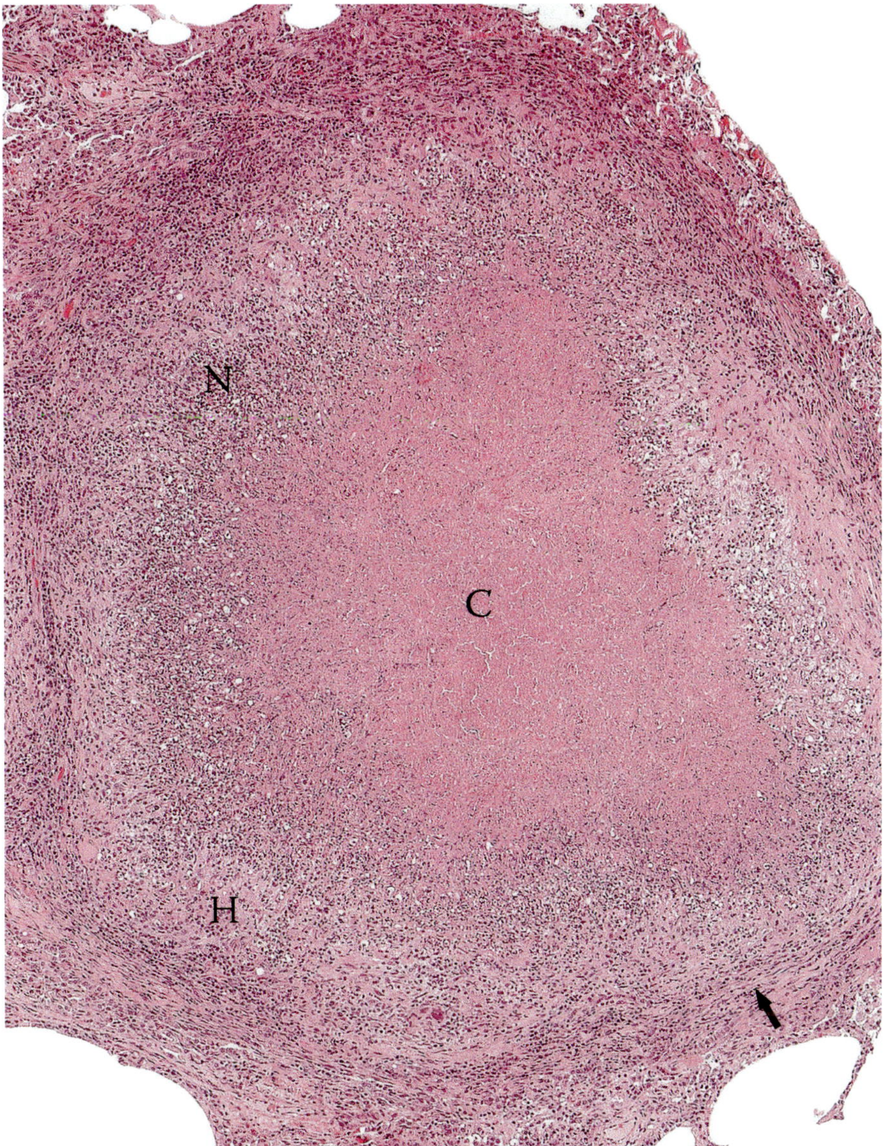

FIGURE 6.4 Necrotizing (caseous) granuloma (H&E, MP). Typical necrotizing granuloma. Note central amorphous eosinophilic debris (C), immediately adjacent margin of degenerative neutrophils and pyknotic nuclear debris (N), more peripheral zone of large epithelioid histiocytes (H), and surrounding thin rim of peripheral fibrosis (arrow). Patchy foci of lymphocytic cells (not abundant in this lesion) are noted by the outer margin.

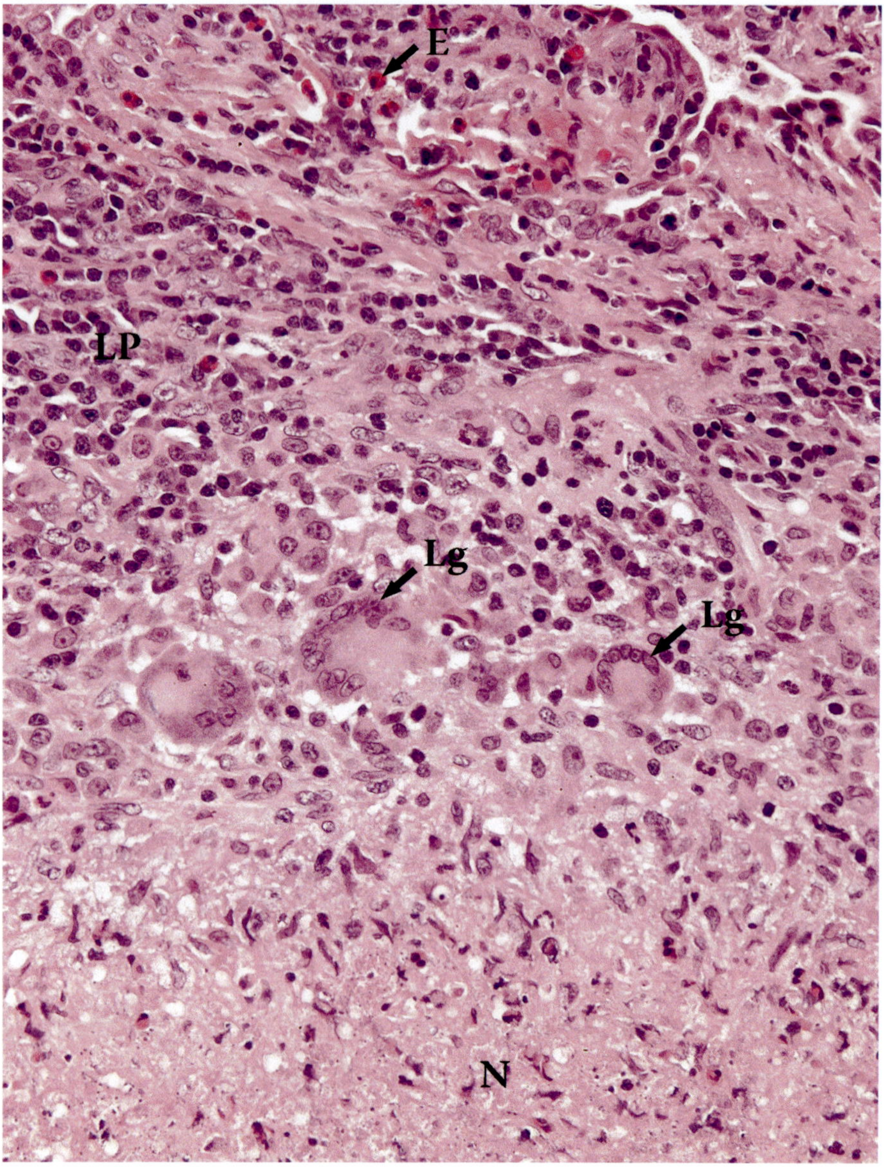

FIGURE 6.5 Necrotizing (caseous) granuloma (H&E, HP). Granuloma architecture is demonstrated in greater detail, including central necrotic (caseous) debris (N), surrounding margin of epithelioid histiocytes with numerous Langhans' type (Lg) and a peripheral mantle of lymphoplasmacytic cells (LP). Note also abundant eosinophils (E) admixed within this outer layer.

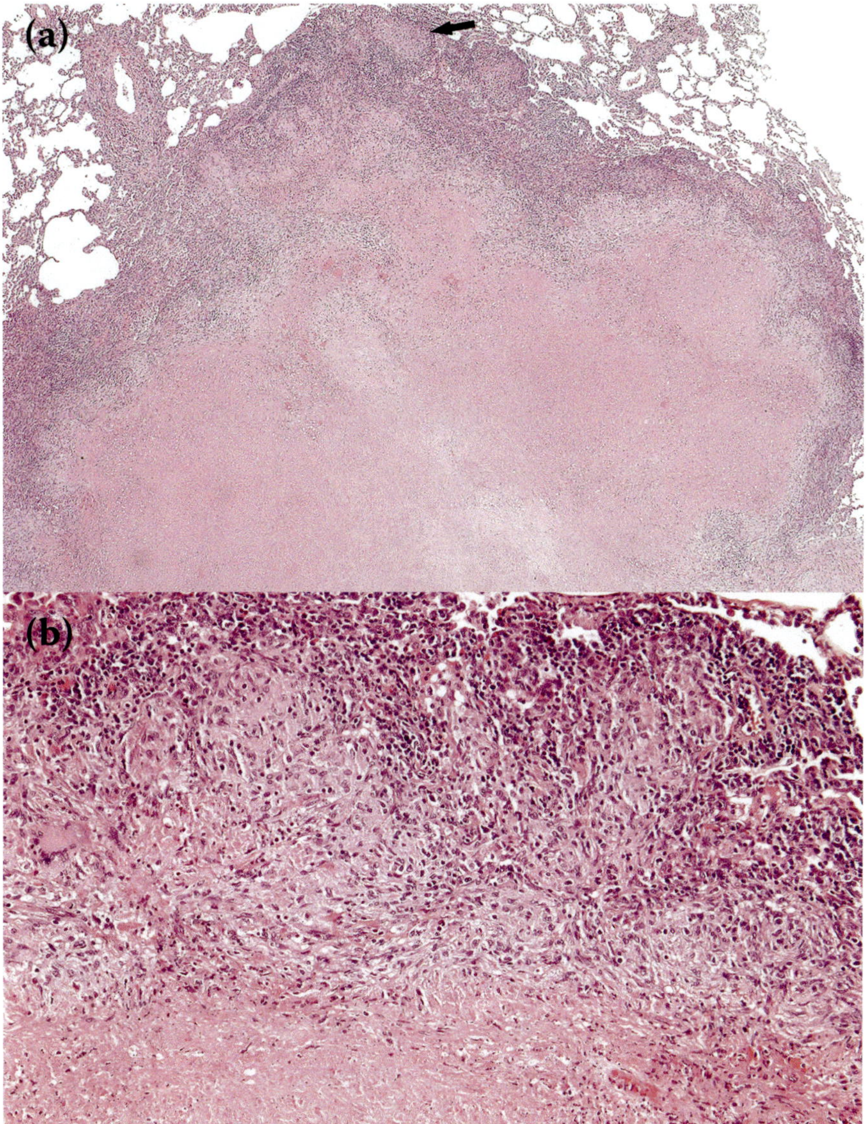

FIGURE 6.6 Local alveolar extension (H&E, LP, and MP). Small aggregates of histiocytes (a, arrow) extend from large underlying necrotizing granulomas into adjacent alveolar areas. (a) LP. (b) MP.

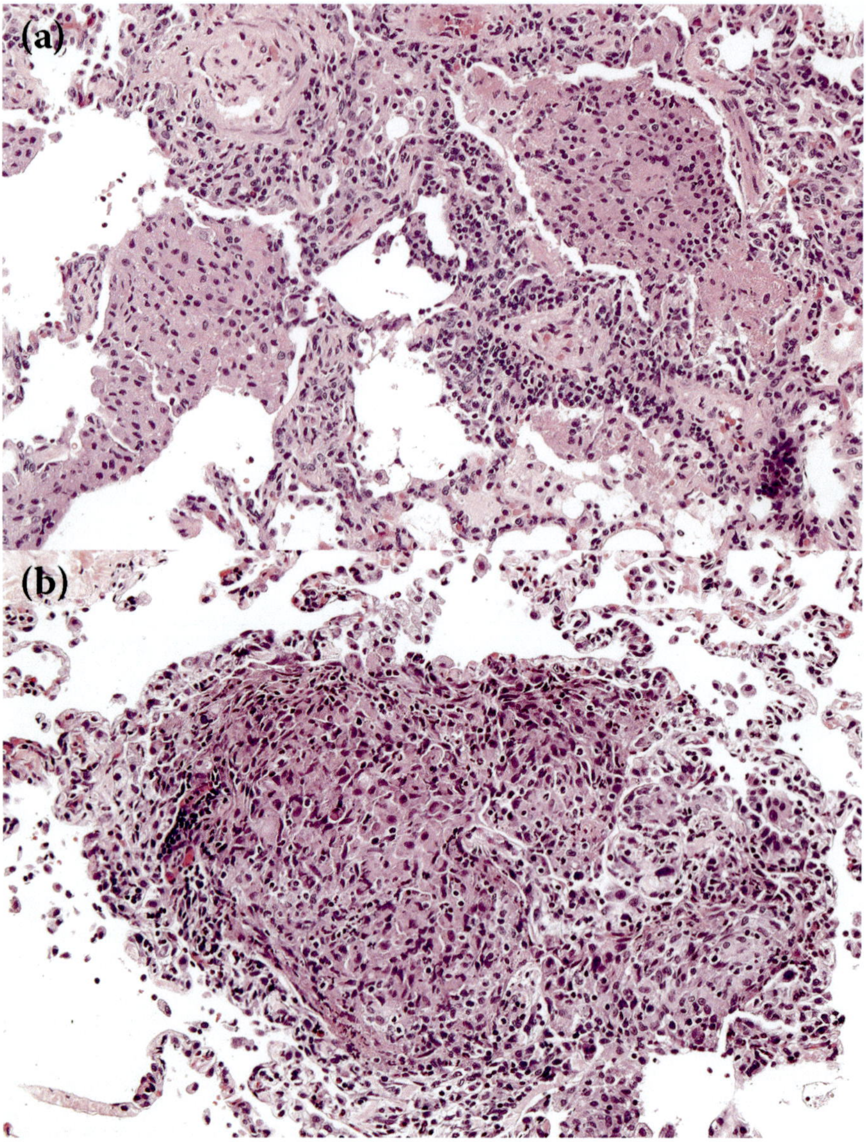

FIGURE 6.7 Alveolar histiocytic foci (a and b; H&E HP). Aggregates of large eosinophilic histiocytic cells present within alveoli independent of adjacent granuloma structures.

fibrous central organization of necrotizing granulomas, with collagen fibers initially forming within the background of necrotic, acellular debris (Figure 6.8a,b), followed by variable levels of more progressive, maturing fibrosis and/or mineralization (Figure 6.2d and Figure 6.8c,d).

Latent

Over half of monkeys experimentally infected with a low dose of MTB develop true latent infection with disease containment (Lin et al., 2009), comparable to humans. In such monkeys, lesions may be subtle and difficult to detect, often being limited to small subpleural or interstitial granuloma scars or fibrocalcific lesions within lung parenchyma and/or associated thoracic lymph nodes. Treatment-healed disease often has a somewhat similar microscopic appearance to latent infection, although the scope of the resolved/resolving lesions is typically much greater due to the degree of lung involvement at the time medication was started. More widespread interstitial scarring is seen, sometimes associated with residual lymphoplasmacytic inflammatory nodules, suggesting persistent antigenic drive (Figure 6.9). Additionally, depending on the stage of therapy, in treatment-healed disease one is more likely to detect a morphological continuum of internal fibrous organization within recognizable granuloma structures, than in latent or spontaneously resolved disease (Figure 6.10).

VARIATIONS

Tuberculous Pneumonia

In poorly contained infection, disease can progress from focal or multifocal inflammatory nodules to larger confluent areas of widespread parenchymal consolidation. Such a process can occur by a coalescence of smaller expanding subunit necrotizing granulomas and/or direct uncontained extension of disease through intact alveolar walls (Figure 6.11). Associated parenchymal infiltration can range from a patchy distribution to complete lobar consolidation.

Cavitation

Cavitation is a process in which liquefaction of central necrotic granuloma content takes place, leading to direct extension of the lesion into an adjacent bronchial airway. The resulting "hollow" space is referred to as a tuberculous cavity. Cavities occurring in active and reactivation tuberculosis consist of a margin of granulation tissue and evolving fibroplasia, with

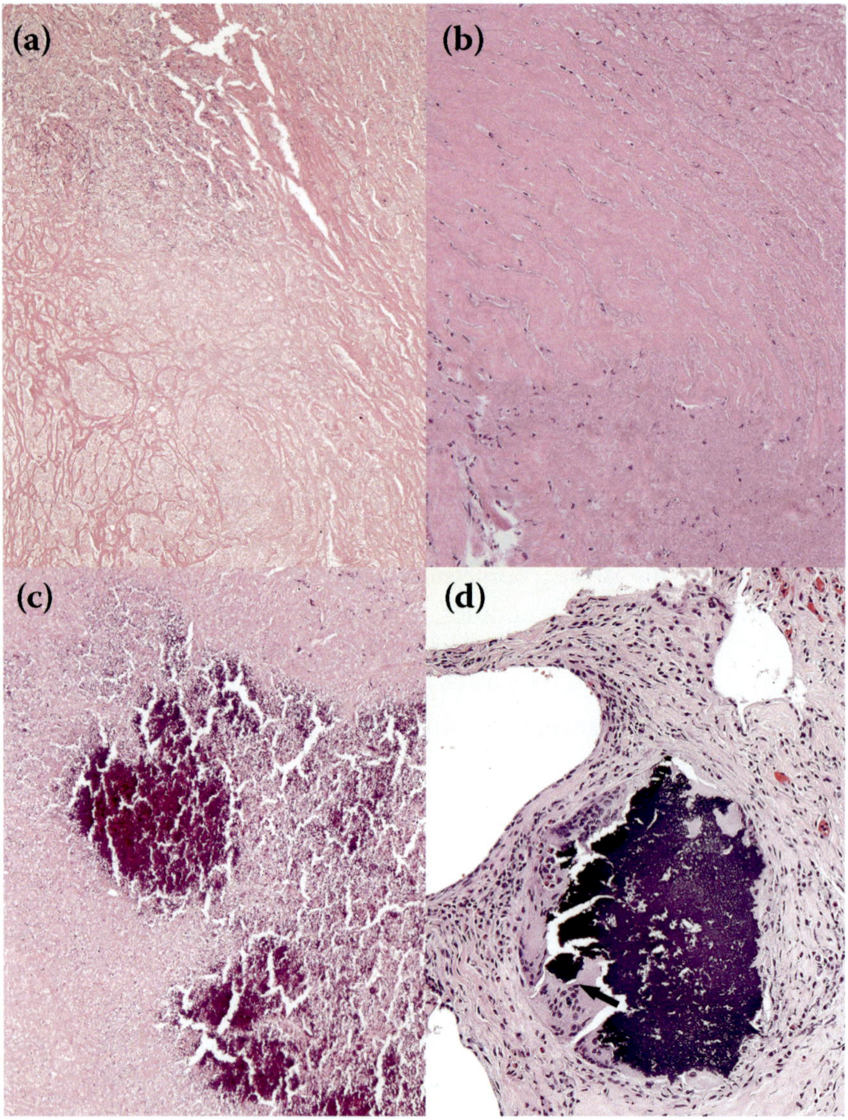

FIGURE 6.8 Collagenous organization and mineralization (calcification). H&E, MP. (a) A network of collagen fibrils is developing within a background of residual debris in this necrotizing granuloma. (b) More dense eosinophilic bands of collagen fibers are evolving in this more chronic necrotizing granuloma. (c) Punctate basophilic foci of mineralization coalescing in areas into large aggregates are present in this necrotizing granuloma. (d) This chronic fibrocalcific lesion is comprised primarily of dense solid mineralized debris. Multinucleated giant cells (arrow) are present as a reaction to the "foreign" calcific material (not a direct response to mycobacterial organisms).

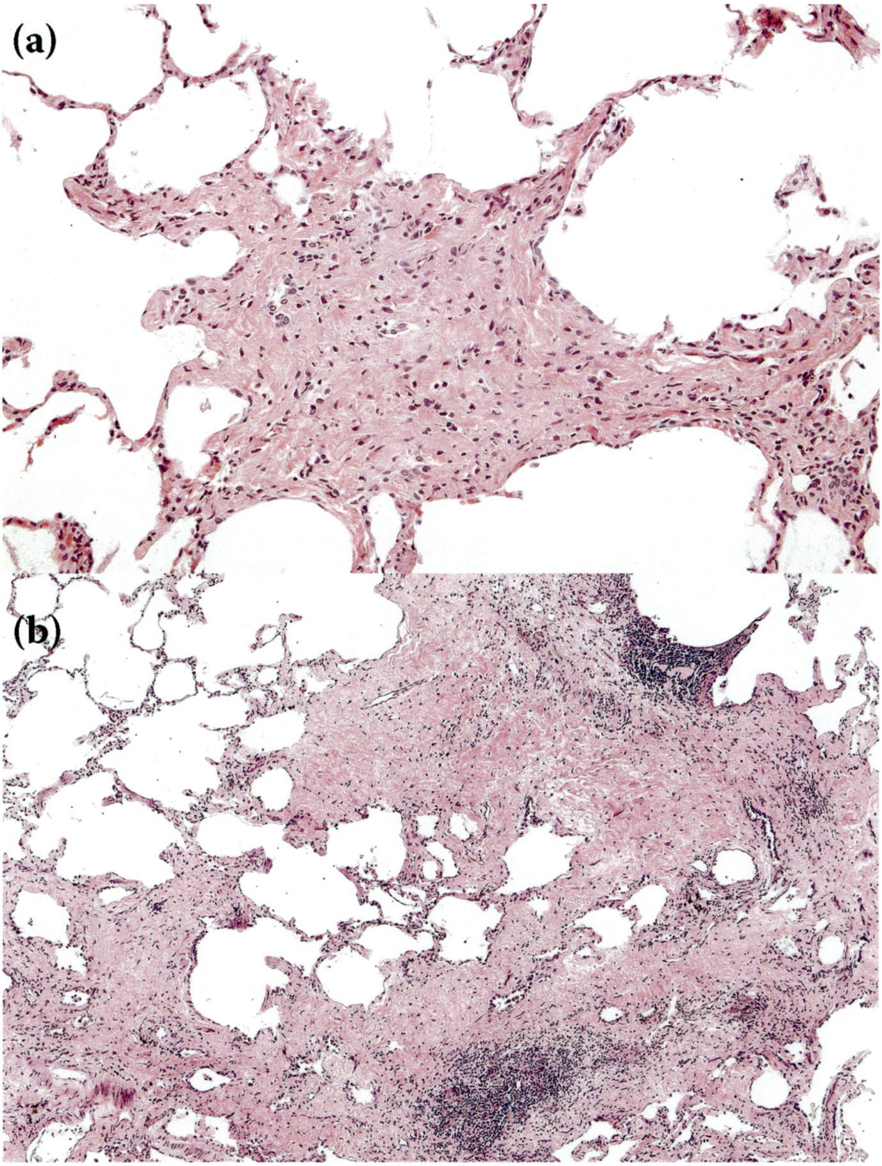

FIGURE 6.9 Interstitial fibrosis/scarring (H&E, MP, and LP). (a) MP. Area of interstitial fibrosis secondary to granuloma resolution (granuloma scar). (b) LP. Interstitial fibrosis (granuloma scar) with patchy adjacent lymphocytic aggregates suggesting persistent antigenic drive.

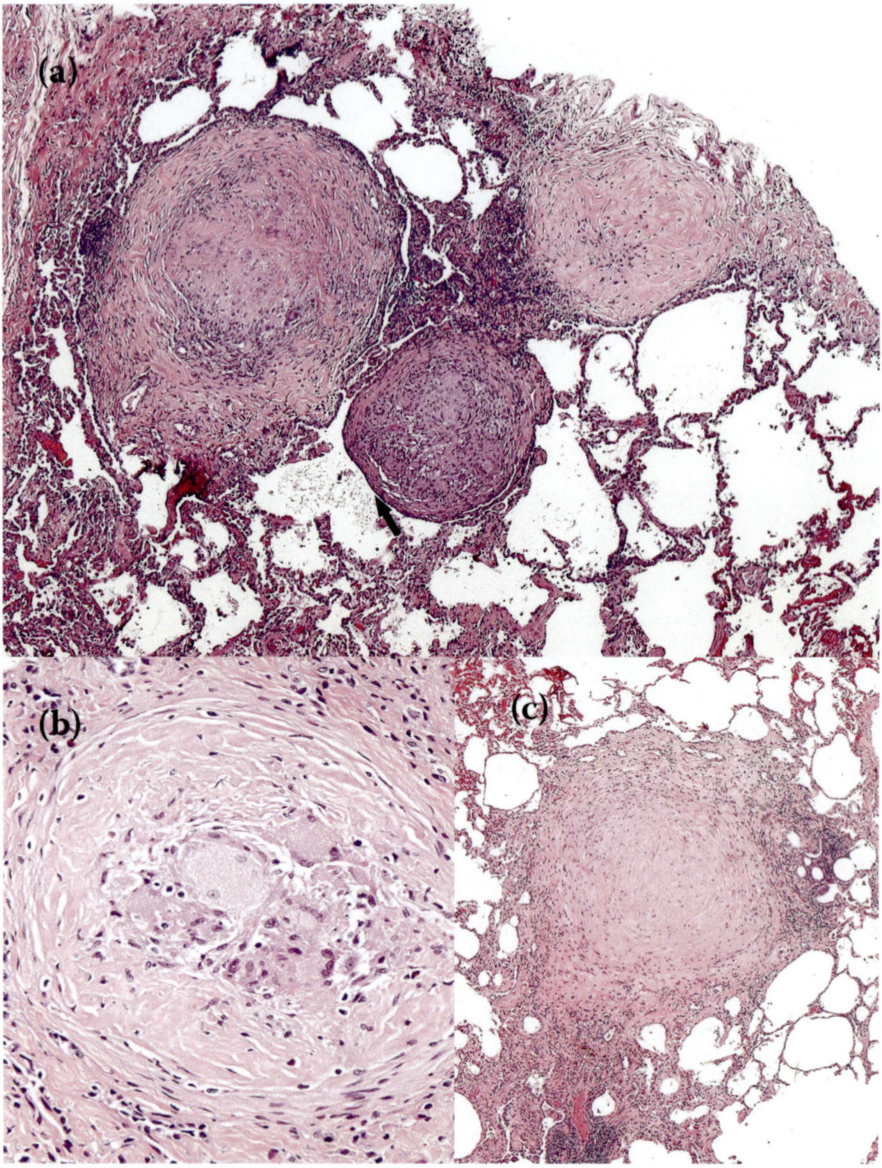

FIGURE 6.10 Healing/fibrous organizing granulomas (H&E). (a) LP. Three separate granulomas, each in a different stage of healing (fibrous organization) due to drug treatment. (b) HP. A granuloma with extensive peripheral fibrous organization and remaining central epithelioid histiocytic population. (c) LP. The content of this granuloma has been completely replaced by fibrous connective tissue.

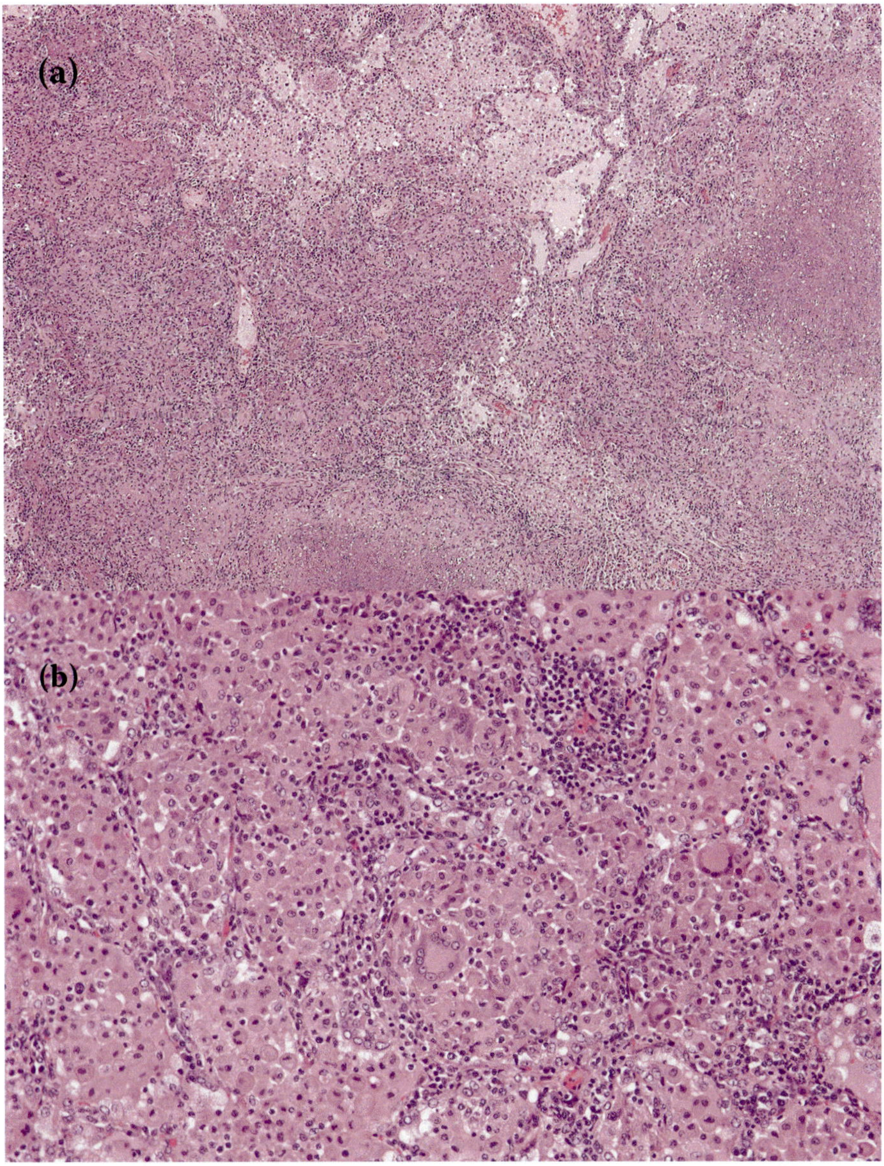

FIGURE 6.11 Tuberculous pneumonia (H&E, LP, and MP). (a) LP. Tuberculous pneumonia characterized by both coalescing caseous granulomas and direct histiocytic infiltration through lung parenchyma. (b) MP. Tuberculous pneumonia characterized by direct extension of epithelioid histiocytes and multinucleated cells through intact alveolar structures without evidence of frank granuloma structure formation.

adherent luminal necrotic debris. Areas of contiguity with invaded bronchi are commonly seen in tissue section (Figure 6.12). Cavities are thought to be the major contributor to TB transmission in humans, and as such, their frequent occurrence in the NHP model is of significant importance.

Suppurative Granulomas

Another variation of granuloma type seen in early, poorly contained NHP tuberculosis is associated with a predominance of central granulocytic inflammation. Although neutrophils are a common component of many necrotizing granulomas, some lesions have a unique and characteristic appearance, being comprised totally of a central core of degenerative polymorphonuclear cells (Figure 6.2c and Figure 6.13). This suppurative variety of granuloma harbors large numbers of organisms, because this liquefactive milieu is an excellent medium for extracellular bacillus growth and replication.

Miliary Disease

The grossly descriptive designation of this disease form, as seen in only a small percentage of both infected humans and monkeys, stems from the gross appearance of the vast number of evenly disseminated and uniformly sized (1 to 2 mm) lesions present—resembling scattered millet seeds. Miliary tuberculosis occurs when there is massive, widespread dissemination of organisms hematogenously. Histologically, these foci although circumscribed, consist primarily of abundant amounts of central amorphous eosinophilic necrotic debris, often with only a modest peripheral inflammatory cellularity (Figure 6.14).

Endobronchial Disease

Endobronchial disease, defined as infection within the tracheobronchial tree, is seen not uncommonly in this model. Lesions range from small microscopic exophytic nodular projections into the bronchial lumen, to larger polypoid structures with extensive mucosal ulceration (Figure 6.15).

Reactivation (Postprimary) Disease

Like people, monkeys experimentally inoculated with MTB can reactivate latent infection at any point, including >3 years after infection. It may be difficult to blindly differentiate primary rapidly progressive infection from the reactivation process, especially if the latter is in an advanced state.

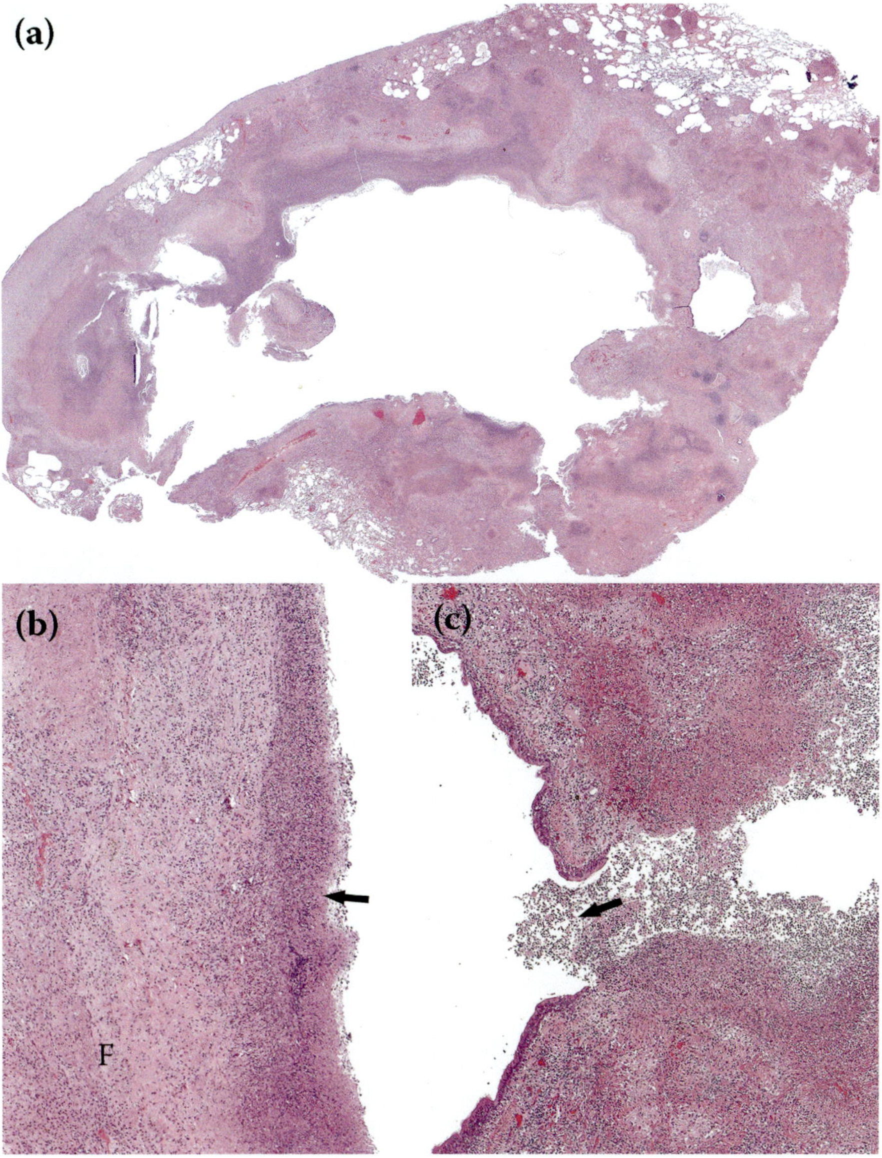

FIGURE 6.12 Cavitation (H&E, LP). (a) Tuberculous cavity. (b) The inner margin of this structure is lined by a combination of residual necrotic debris and degenerative neutrophils (arrow), while evolving fibrosis is noted more peripherally (F). (c) Point of communication between a tuberculous cavitary lesion and an adjacent bronchus (arrow).

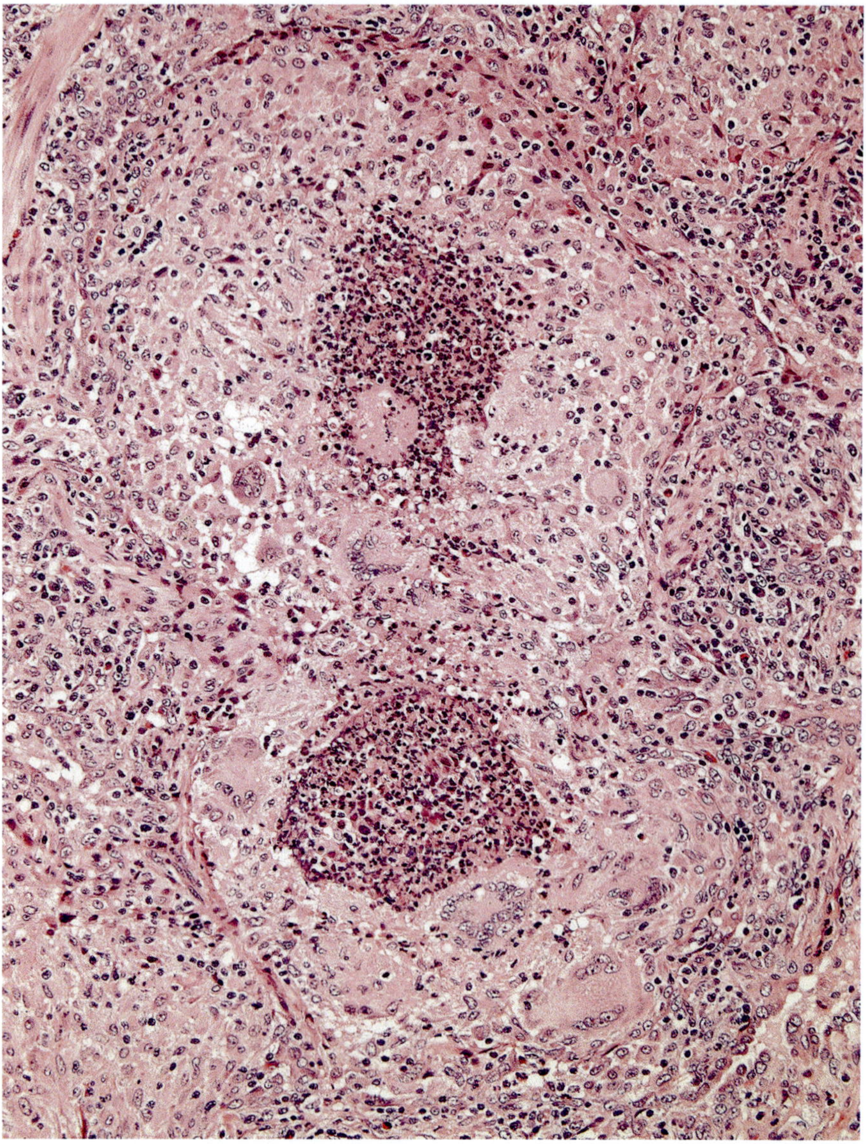

FIGURE 6.13 Suppurative granuloma (H&E, HP). Two coalescing suppurative granulomas with central cores of neutrophils surrounded by large epithelioid histiocytes and multinucleated giant cells.

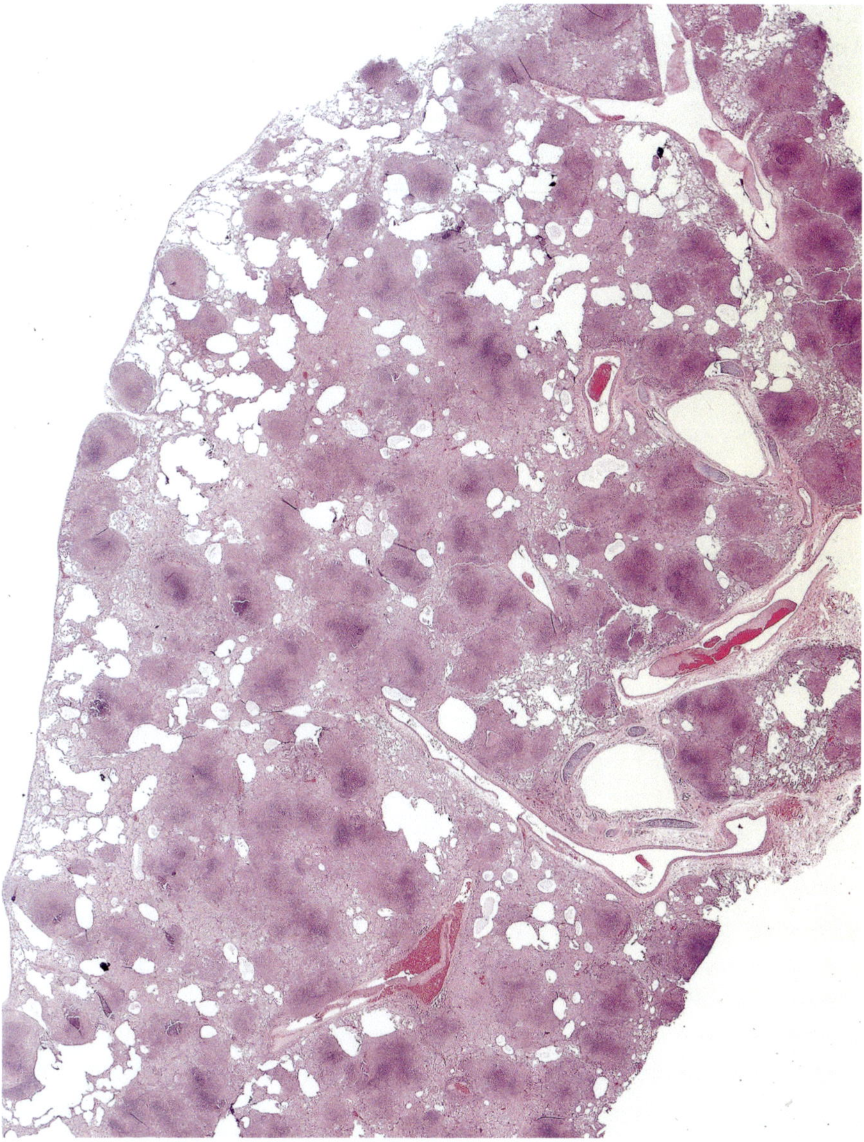

FIGURE 6.14 Miliary tuberculosis (H&E, LP). Multifocal and coalescing uniformly sized necrotizing granulomas reflecting hematogenous disease dissemination.

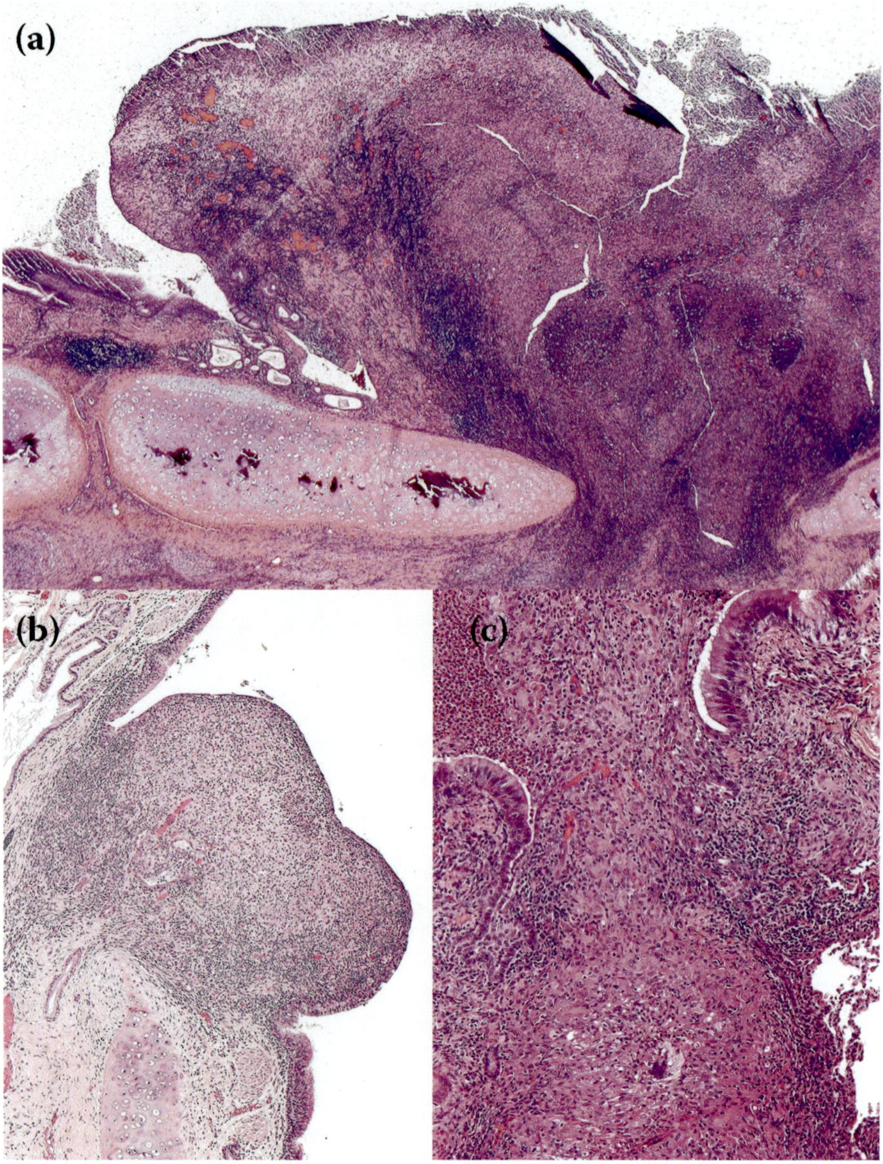

FIGURE 6.15 Endobronchial disease (H&E). (a) LP. A large, somewhat polypoid endobronchial mass comprised of necrotizing granulomatous inflammation with prominent admixed lymphocytic component extends into the lumen of this large airway. (b) LP. A smaller exophytic endobronchial nodule infiltrates into this bronchus, leading to a focally extensive area of mucosal ulceration. (c) MP. Direct bronchial invasion by an adjacent necrotizing granuloma.

Histologically, perception of obviously chronic lesions as characterized by dense, maturing fibrosis and/or extensive mineralization in conjunction with more acute, active processes, such as the presence of numerous nonnecrotizing granulomas, tuberculous pneumonia, or alveolar histiocytic foci can strongly suggest that a reactivation process has occurred (Figure 6.16).

STRENGTHS OF THIS ANIMAL MODEL

There are two major strengths of the macaque model: (1) The pathology and varieties of granulomas are remarkably similar to that seen in humans. (2) The outcomes of infection, from latent to chronic to active to fulminant disease are extremely similar to the human spectrum of disease. Most extrapulmonary manifestations reported in humans, including Pott's disease (skeletal TB), miliary TB, spleen and liver involvement, meningitis and cerebral TB, cardiac TB, and ocular TB, have been observed in the macaque model. The availability of reagents combined with the similarities of the disease to humans makes this an ideal model for the study of tuberculosis.

SHORTCOMINGS

The difficulties in working with nonhuman primates in a Biosafety Level 3 facility complicate the use of this model. Most BSL-3 facilities do not have housing for MTB-infected monkeys or are limited by space. Nonhuman primates require special husbandry and enrichment. The cost of performing studies in nonhuman primates is high, including purchasing the animals, housing, and procedures. Trained veterinarians and technicians must be an essential part of the research team. Finally, macaques are outbred, which greatly increases the variability among animals. Although this is a benefit for reproducing the variability seen in humans infected with MTB and, of course, contributes to the spectrum of tuberculosis in macaques, the variability increases the numbers of animals needed for statistically significant results.

REFERENCES

Capuano, S.V., 3rd, Croix, D.A., Pawar, S., Zinovik, A., Myers, A., Lin, P.L., Bissel, S., Fuhrman, C., Klein, E., and Flynn, J.L. 2003. Experimental *Mycobacterium tuberculosis* infection of cynomolgus macaques closely resembles the various manifestations of human *M. tuberculosis* infection. *Infect Immun,* 71, 5831–5844.

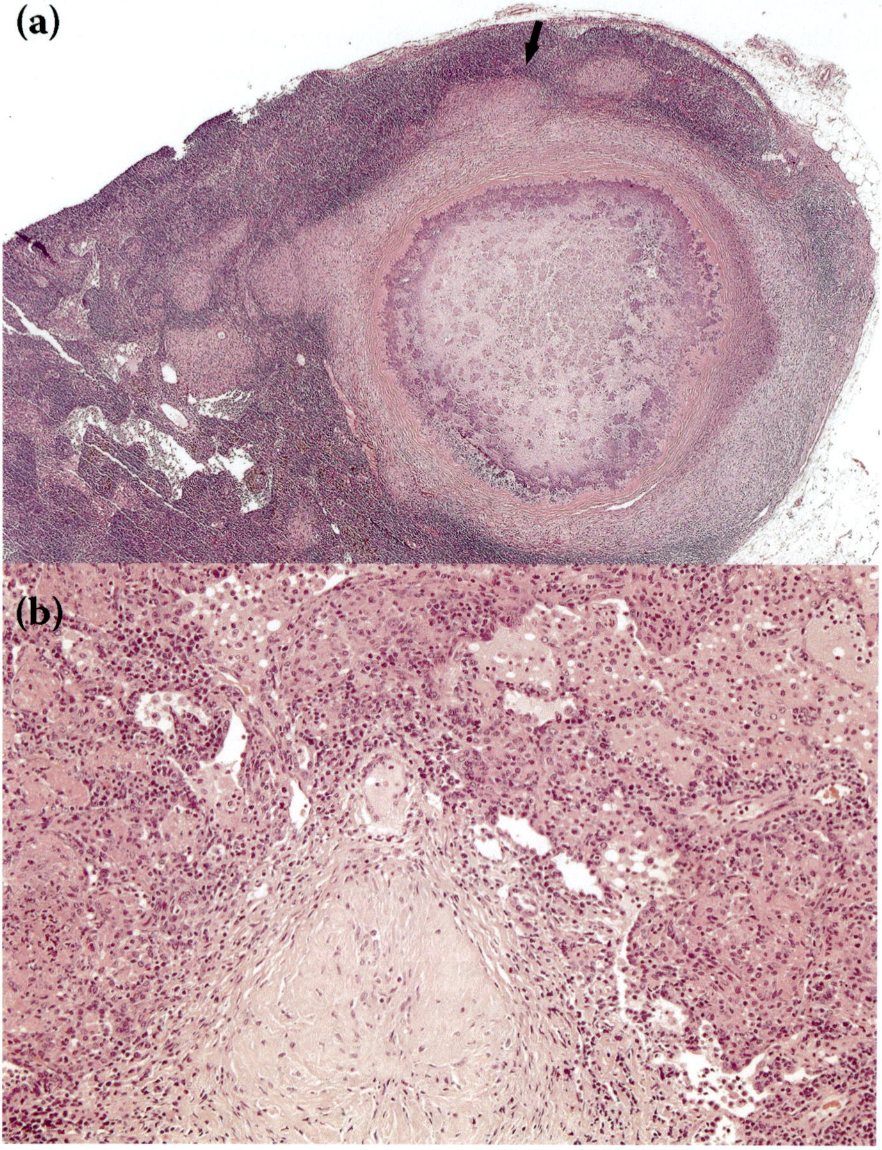

FIGURE 6.16 Reactivation disease (H&E, LP, and HP). (a) Thoracic lymph node draining an area of tuberculous infected lung. A large chronic fibrocalcific granuloma is present eccentrically. Also noted are multiple foci of nonnecrotizing histiocytic inflammation, including several extending directly out of the larger chronic lesion (arrow). (b) Reactivation disease characterized by the concurrent presence of a chronic sclerotic granuloma and adjacent active tuberculous pneumonia.

Chen, C.Y., Huang, D., Wang, R.C., Shen, L., Zeng, G., Yao, S., Shen, Y., Halliday, L., Fortman, J., McAllister, M., Estep, J., Hunt, R., Vasconcelos, D., Du, G., Porcelli, S.A., Larsen, M.H., Jacobs, W.R., Jr., Haynes, B.F., Letvin, N.L., and Chen, Z.W. 2009. A critical role for CD8 T cells in a nonhuman primate model of tuberculosis. *PLoS Pathog,* 5, e1000392.

Flynn, J.L., Cooper, A., and Bishai, W.R. 2005. Animal models of tuberculosis. In *Tuberculosis and the tubercle bacillus,* eds. S.T. Cole, K.D. Eisenach, D.N. McMurray, and W.R. Jacobs, Jr. Washington, DC: ASM Press, 547–560.

Good, R.C. 1968. Simian tuberculosis: Immunologic aspects. *Ann NY Acad Sci,* 154, 200–213.

Langermans, J.A., Doherty, T.M., Vervenne, R.A., Van Der Laan, T., Lyashchenko, K., Greenwald, R., Agger, E.M., Aagaard, C., Weiler, H., Van Soolingen, D., Dalemans, W., Thomas, A.W., and Andersen, P. 2005. Protection of macaques against *Mycobacterium tuberculosis* infection by a subunit vaccine based on a fusion protein of antigen 85B and ESAT-6. *Vaccine,* 23, 2740–2750.

Langermans, J.A.M., Andersen, P., Van Soolingen, D., Vervenne, R.A.W., Frost, P.A., Van Der Laan, T., Van Pinsteren, L.A.H., Van Den Hombergh, J., Kroom, S., Peekel, I., Florquin, S., and Thomas, A.W. 2001. Divergent effect of bacillus Calmette-Guerin (BCG) vaccination on *Mycobacterium tuberculosis* infection in highly related macaque species: Implications for primate models in tuberculosis vaccine research. *Proc Nat Acad Sci USA,* 98, 11497–11502.

Lin, P.L., Pawar, S., Myers, A., Pegu, A., Fuhrman, C., Reinhart, T.A., Capuano, S.V., Klein, E., and Flynn, J.L. 2006. Early events in *Mycobacterium tuberculosis* infection in cynomolgus macaques. *Infect Immun,* 74, 3790–3803.

Lin, P.L., Rodgers, M., Smith, L., Bigbee, M., Myers, A., Bigbee, C., Chiosea, I., Capuano, S.V., Fuhrman, C., Klein, E., and Flynn, J.L. 2009. Quantitative comparison of active and latent tuberculosis in the cynomolgus macaque model. *Infect Immun,* 77, 4631–4642.

Verreck, F.A., Vervenne, R.A., Kondova, I., Van Kralingen, K.W., Remarque, E.J., Braskamp, G., Van Der Werff, N.M., Kersbergen, A., Ottenhoff, T.H., Heidt, P.J., Gilbert, S.C., Gicquel, B., Hill, A.V., Martin, C., McShane, H., and Thomas, A.W. 2009. MVA.85A boosting of BCG and an attenuated, phoP deficient M. tuberculosis vaccine both show protective efficacy against tuberculosis in rhesus macaques. *PLoS ONE,* 4, e5264.

CHAPTER 7

Pulmonary Tuberculosis in the Rabbit

Gilla Kaplan and Liana Tsenova

CONTENTS

INTRODUCTION

The rabbit was first used by Medlar and Sasano as a model for tuberculosis in 1935. These authors compared the lung lesions induced in response to infection with three differently virulent strains of *Mycobacterium bovis* BCG. They described "a variety of (lung) lesions … similar to those, known to occur in the human tuberculosis lung" (Medlar and Sasano, 1935). The lesions were characterized as showing "localization, progression, cavitation, bronchogenic spread and retrogression." These studies suggested that the type of disease induced in infected rabbits is a function of the balance between the resistance of the host and the virulence of the pathogen. Max B. Lurie, and later Lurie and Dannenberg, extensively characterized the rabbit response to infection with *M. bovis* Ravenel or *Mycobacterium tuberculosis* (MTB) H37Rv, drawing many analogies to the progression of human disease (Lurie, 1928, 1932; Lurie et al., 1950, 1952; Dannenberg, 1994; Converse et al., 1998). Depending on the dose of infecting organisms and the mycobacterial strain, Lurie and Dannenberg observed either mild infection that was controlled by the rabbit immune response (usually caused by MTB H37Rv infection), or severe progressive disease, often with formation of necrotizing granulomas with and without cavities (usually caused by *M. bovis* Ravenel infection). Consequently, rabbits were considered relatively resistant to MTB compared to *M. bovis* Ravenel, suggesting the development of a robust protective acquired immune response to infection with the human pathogen (Lurie, 1964; Dannenberg, 2001, 2006). Taken together, these studies demonstrated that the rabbit can be used as an excellent model for tuberculosis infection, mimicking all forms of human disease.

INOCULUM AND ROUTE OF INOCULATION

We have established a protocol for use of a nose-only aerosol exposure system (CH Technologies, Inc., Westwood, NJ) for studying experimental pulmonary tuberculosis in rabbits. To evaluate and compare the kinetics of infection and progression of disease with strains of different virulence, we used MTB HN878 [more virulent in the mouse (Manca et al., 2001)] or CDC1551 [less virulent in the mouse (Manca et al., 1999)]. New Zealand White rabbits were infected using a suspension of 2×10^7 CFU/ml and a constant aerosol exposure time of 20 minutes. Three hours after infection, subsets of rabbits were euthanized to evaluate the effectiveness of aerosol delivery. By sampling multiple areas of the lung, we observed that approximately 3.6 $\log_{10}$ organisms were deposited evenly throughout the lungs. Groups of rabbits were euthanized at 4, 8, 12, and 16 weeks, and clinical and histopathology parameters were evaluated. At each time point, tissue samples were collected from the lungs, liver, spleen, brain, and selected lymph nodes, and bacillary loads were evaluated by the CFU assay.

FEATURES

Bacillary Load in the Lungs of Infected Rabbits

Although the inoculum for both MTB HN878 and CDC1551 was similar at 3 hours postexposure, their differential virulence was clearly seen by 4 weeks postinfection: MTB HN878 reached 6.9 $\log_{10}$ CFU, whereas CDC1551 reached 5.5 $\log_{10}$. While infection with HN878 resulted in a sustained high bacillary load in the lungs (>6 $\log_{10}$) throughout the experiment, growth of CDC1551 in the lungs reached a plateau at 4 weeks, and over the next 12 weeks of infection, declined, resulting in very few to no culturable bacilli by 16 weeks postinfection (Flynn et al., 2008). Interestingly, HN878 disseminated to the liver and spleen of the majority of infected rabbits. However, no dissemination of CDC1551 to the liver or spleen was noted at any time postinfection. Our results confirmed the earlier observations of Lurie and more recently of Manabe et al., demonstrating differential virulence of mycobacterial strains in the rabbit lung (Lurie et al., 1950; Lurie, 1964; Manabe et al., 2003).

Gross Pathology

Examination of the lungs of rabbits infected with HN878 or CDC1551 at 4 weeks postinfection revealed no clearly visible granulomas in any of the rabbits. However, at 8 weeks postinfection with HN878, multiple 2- to 5-mm lesions were visible. By 16 weeks post HN878 infection, some of the granulomas had increased in size, while others appeared to undergo resorption. A small subset of lesions showed extensive enlargement (11 to 16 mm) with cavity formation. These highly enlarged lesions usually localize in the upper apical regions of the lungs. Thus, infection of rabbits with MTB HN878 appeared to give rise to progressive chronic granulomatous disease. In contrast, CDC1551 was controlled and the lesions and bacilli cleared from rabbit lungs, possibly establishing a state of latency.

Histopathology

The principal structural features of the lower respiratory tract in the rabbit are very similar to those in humans (Figure 7.1).

Examination of lung sections from rabbits infected with CDC1551 or HN878 showed similar histopathology at 2 weeks postinfection. Both were characterized by increased cellularity in the alveolar spaces with reactive peribronchial and perivascular lymph nodes (not shown). However, from 4 weeks postinfection the two strains gave rise to different disease patterns. At 4 weeks postinfection, sections of lungs from rabbits infected with CDC1551 showed few very small cellular aggregates (Figure 7.2a). The granulomas increased in size by 8 weeks (Figure 7.2b and Table 7.1) and became well organized, with a central area of histiocytes and a peripheral cuff of lymphocytes (Figure 7.3). By 12 weeks few small well-differentiated granulomas with no acid-fast bacilli were seen (Figure 7.4). By 16 weeks the lesions had undergone resorption to a point where only increased cellularity was observed in the parenchyma (not shown). Only a single small

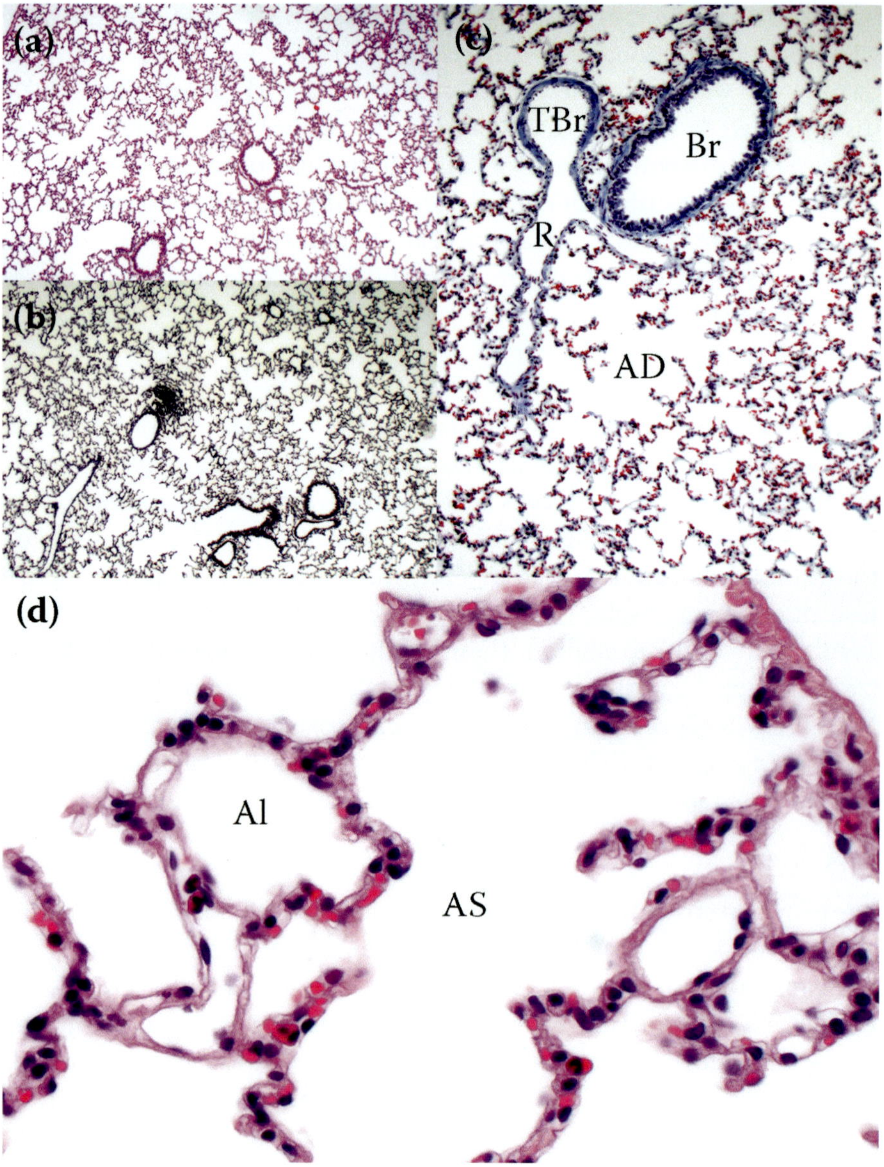

FIGURE 7.1 Normal structure of lower respiratory tract in the rabbit. (a) H&E, LP, (b) Van Gieson's stain, LP, (c) trichrome stain, MP, (d) H&E, HP. The terminal portion of the respiratory tree contains bronchioles (Br), terminal bronchioles (TBr) that branch into transitional airways, the respiratory bronchioles (R), and alveolar ducts (AD). These finally terminate in dilated spaces, called alveolar sacs (AS) and alveoli (Al) seen here below the pleura.

TABLE 7.1 Morphometry of the Granulomas in Rabbit Lungs

	CDC1551		HN878	
Time Postinfection (weeks)	**Area of Granuloma (mm^2)**	**No. of Granulomas per cm^2**	**Area of Granuloma (mm^2)**	**No. of Granulomas per cm^2**
4	0.16[a]	0.4 ± 0.4[c]	3.07 ± 0.06	3.2 ± 0.4
8	0.50 ± 0.22[b]	0.6 ± 0.9[d]	4.27 ± 1.90	6.2 ± 1.4
12	ND	ND	2.05 ± 0.16	5.8 ± 1.2

[a] $P = 0.001$.
[b] $P < 0.001$.
[c] $P = 0.007$.
[d] $P < 0.001$.

necrotic lesion was found in the lungs of one of four rabbits infected with CDC1551 for 16 weeks.

In contrast, at 4 weeks and even more so at 8 weeks, the lungs of HN878-infected rabbits contained multiple lesions (Figure 7.5 and Table 7.1). At 4 weeks, perivascular and peribronchiolar reactive lymphoid aggregates were seen (Figure 7.6). The granulomas consisted of scattered lymphocytes, histiocytes, and polymorphonuclear cells (PMN; these cells in the rabbit contain red granules when stained with H&E) (Figure 7.7). By 8 weeks postinfection, some of the lesions became confluent, surrounded by extensive emphysema (Figure 7.5b). Multinucleated giant Langhans' cells were often seen in the areas rich in histiocytes (Figure 7.5 insert and Figure 7.8). Central necrosis was noted in most granulomas from 8 weeks postinfection (Figure 7.8). At this time, the lesions occupied more than 20% of the lung parenchyma (Table 7.1). Many acid-fast bacilli were seen within the histiocytes and in the central necrotic areas of the granulomas (Figure 7.9). At 12 weeks postinfection the extent of the central necrosis in the granulomas had increased (Figure 7.10a), and liquefaction was observed in the center of a subset of the granulomas (suppurative granulomas) (Figure 7.11). This feature is common for tuberculosis in rabbit and human lungs. By 16 weeks, some of the granulomas had enlarged and formed cavities. Others had reduced in size, and mineralization (calcification) could be seen (Figure 7.10b and Figure 7.12a). Calcified granulomas had very few if any acid-fast bacilli (Figure 7.12b). The majority of lesions had disappeared (resorbed).

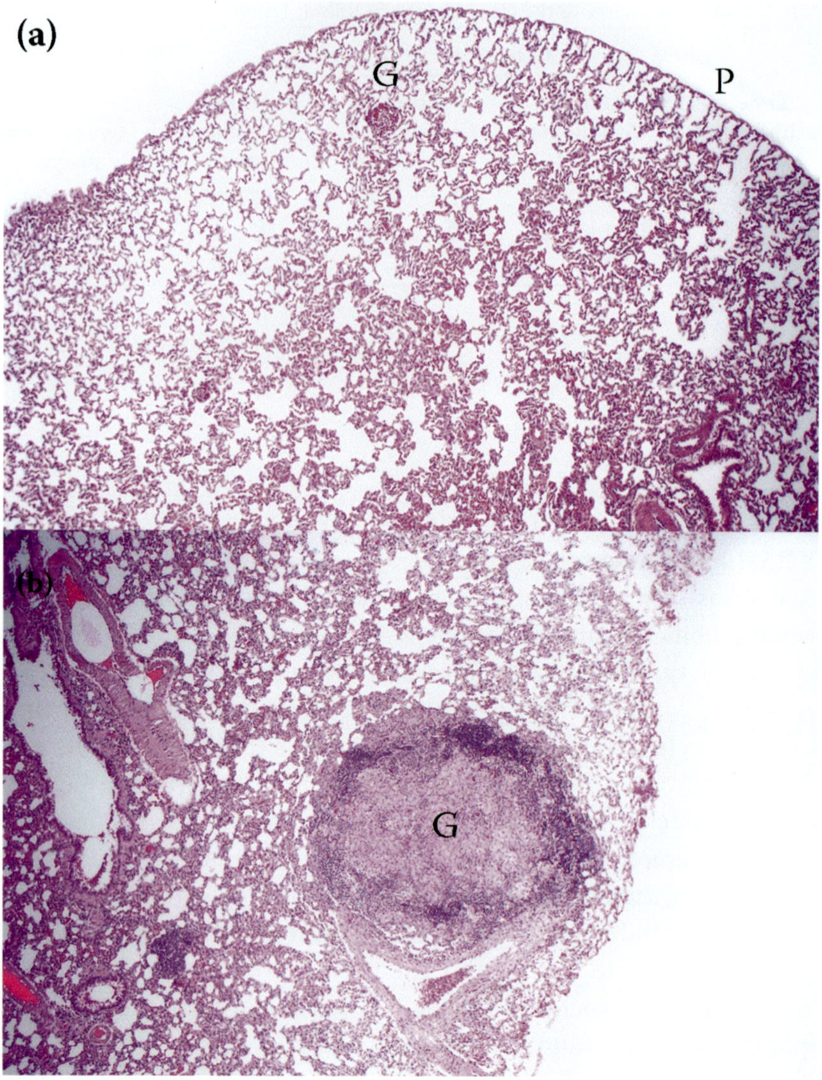

FIGURE 7.2 Infection with CDC1551 (H&E, LP). (a) LP. Four weeks postinfection. Few, very small granulomas (G). P—pleural surface. (b) MP. Eight weeks postinfection. Larger granulomas are seen (G).

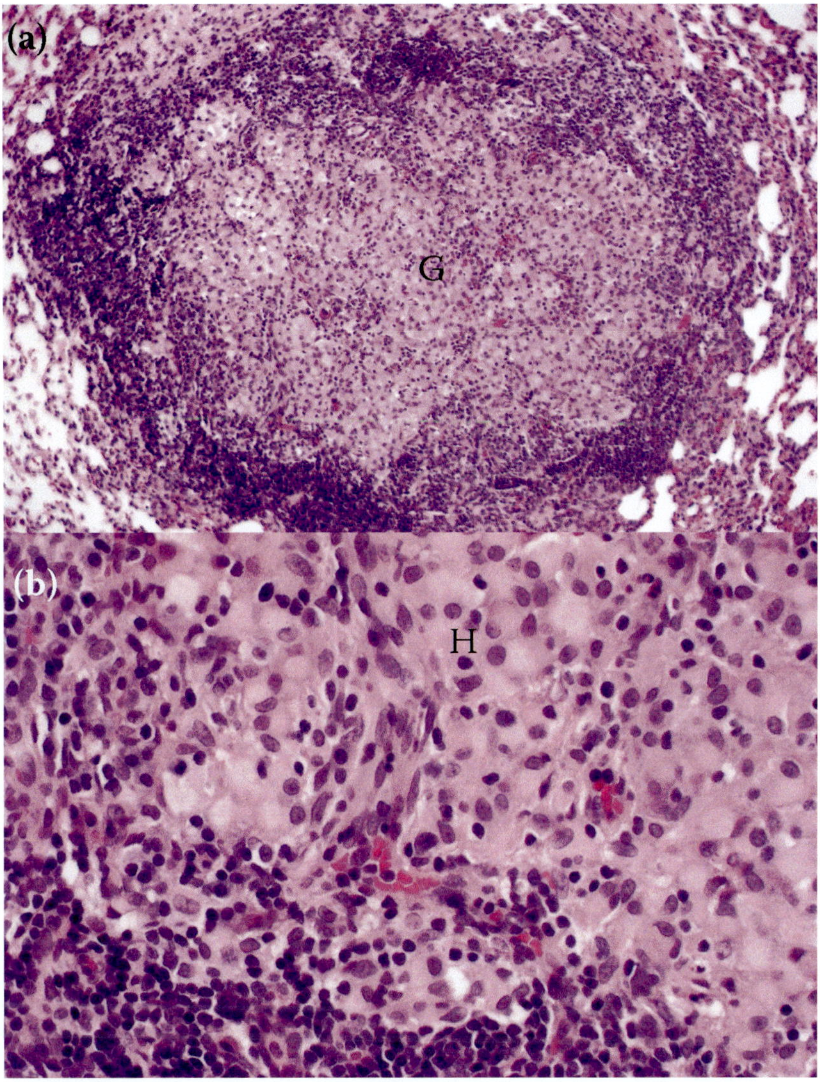

FIGURE 7.3 Infection with CDC1551, 8 weeks postinfection (H&E, MP, and HP). Well-organized granulomas, with a central area of histiocytes (H) and a peripheral cuff of lymphocytes. (a) MP. (b) HP.

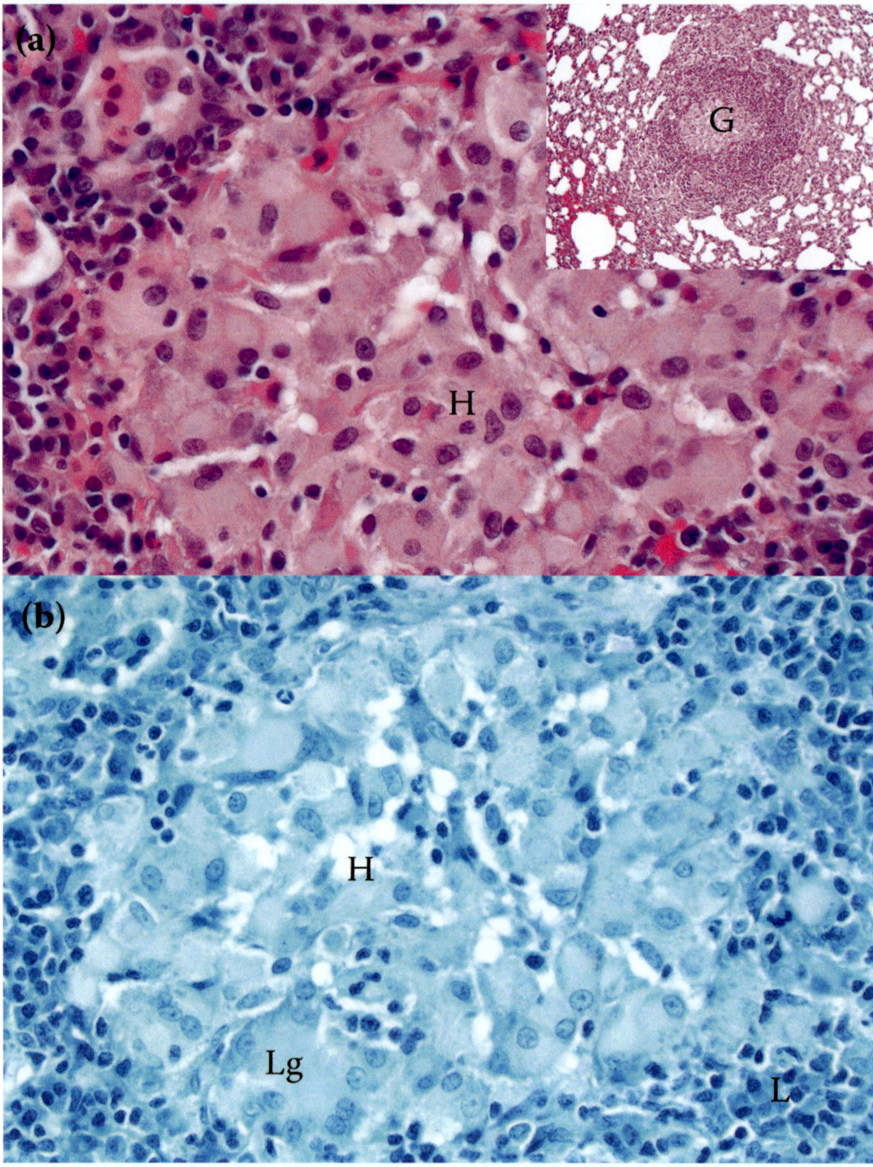

FIGURE 7.4 Infection with CDC1551, 12 weeks postinfection (HP). Small well-defined granuloma (LP inset image) with no acid-fast bacilli. (a) H&E, HP. (b) ZN, HP. G—granuloma. H—epithelioid histiocytes. L—lymphocytes. Lg—Langhans' type multinucleate giant cell.

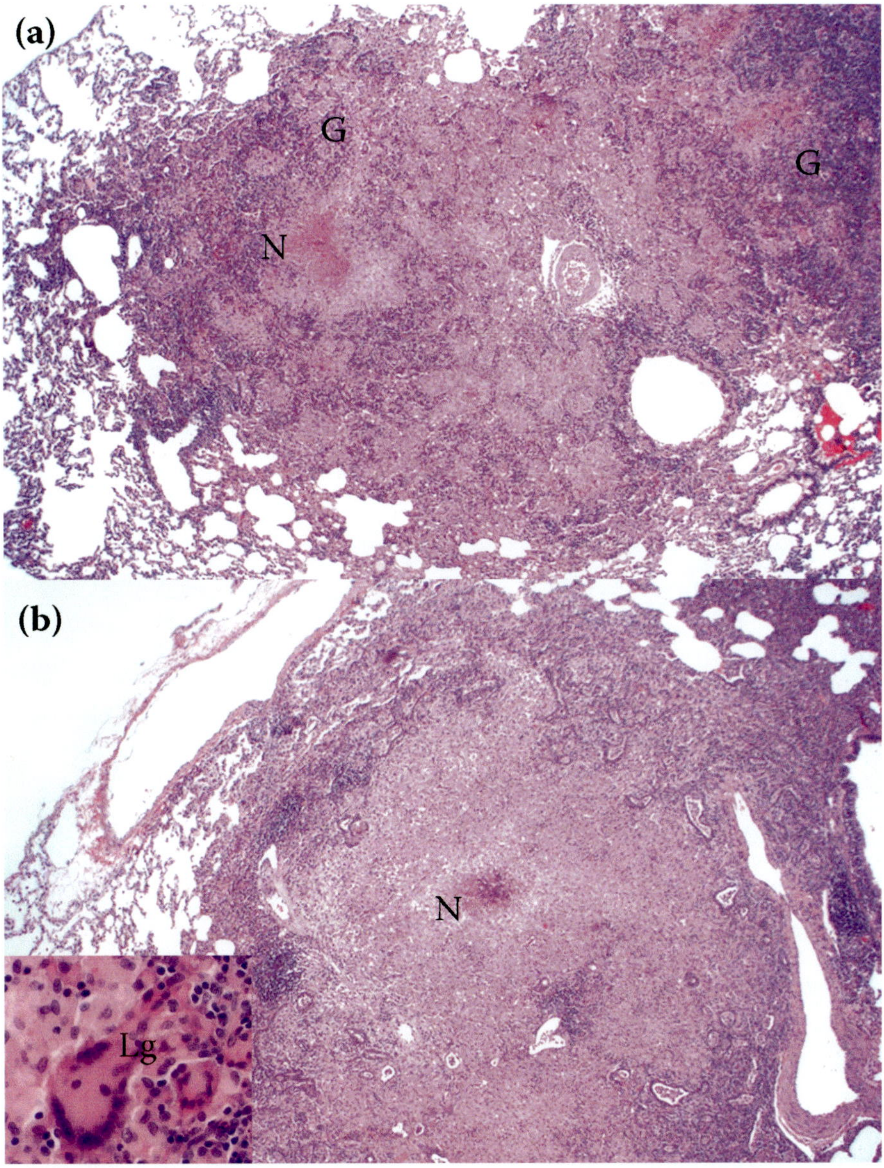

FIGURE 7.5 Infection with HN878 (H&E, LP). Multiple large lesions with central necrosis (N) and surrounding macrophages. By 8 weeks, foamy macrophages (foamy histiocytes) can be seen. (a) 4 weeks postinfection, (b) 8 weeks postinfection; insert—Langhans' type giant cells (Lg).

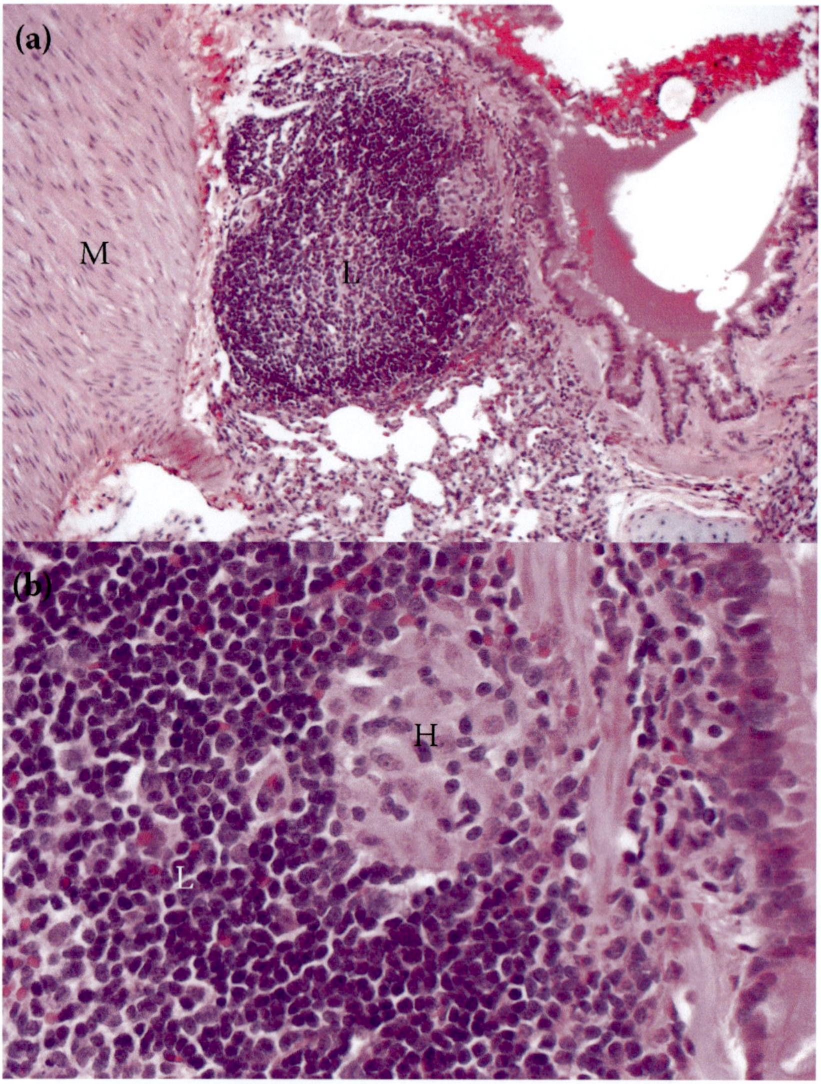

FIGURE 7.6 Infection with HN878, 4 weeks postinfection (H&E, MP, and HP). Perivascular and peribronchiolar reactive lymphoid aggregates (L). H—epithelioid histiocytes. M—smooth muscle. (a) MP. (b) HP.

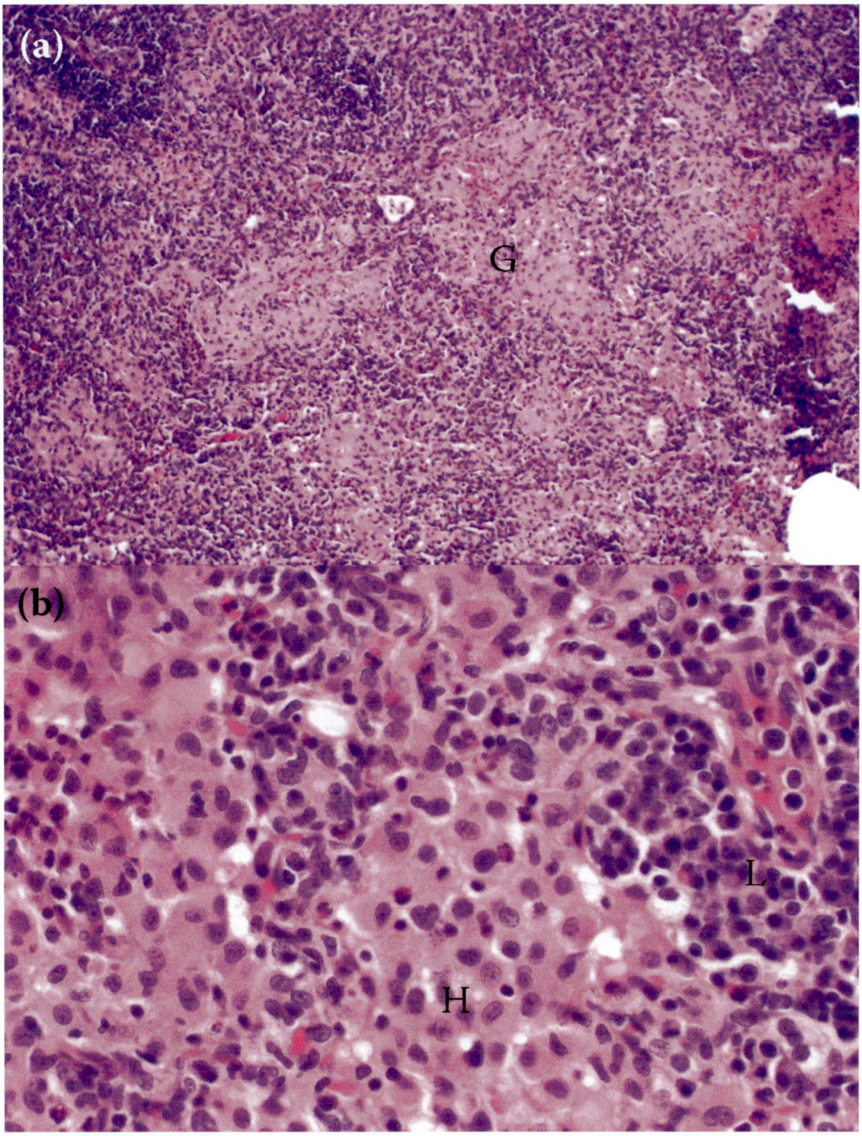

FIGURE 7.7 Infection with HN878, 4 weeks postinfection (H&E). The granulomas (G) contain lymphocytes (L) and histiocytes (H). (a) MP. (b) HP.

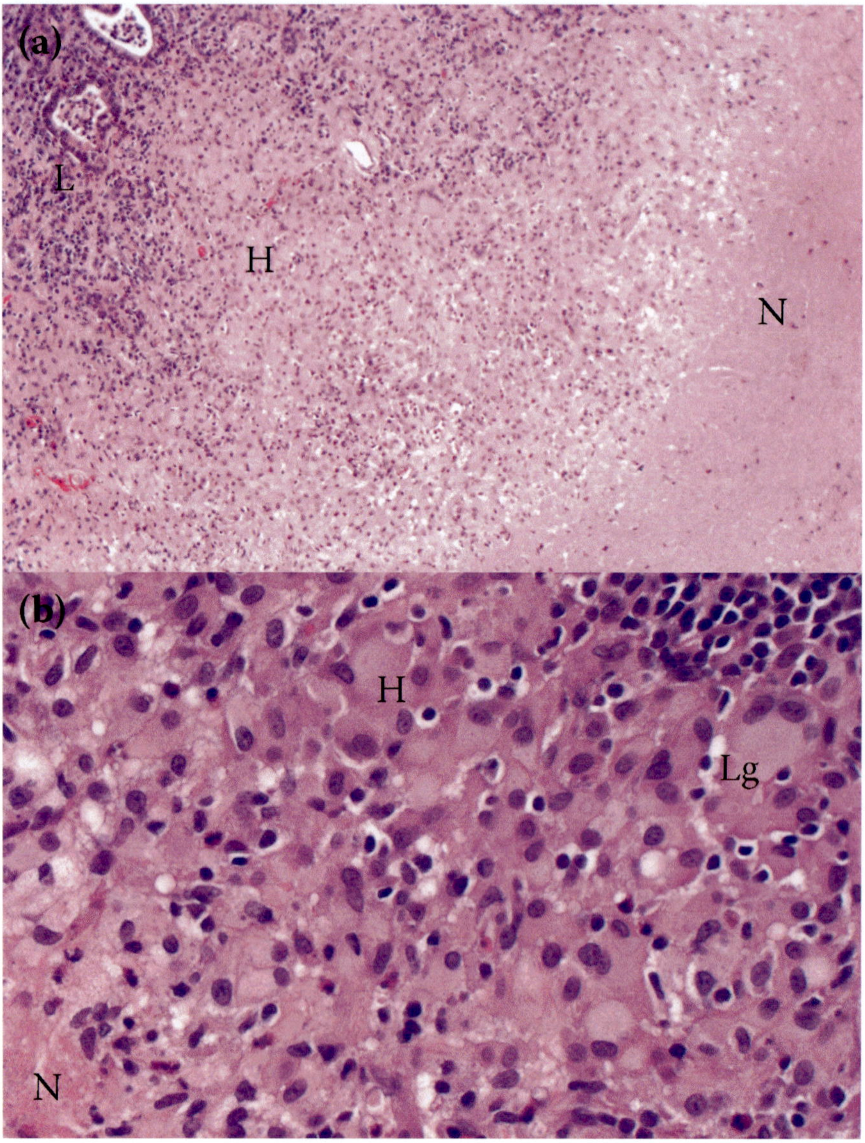

FIGURE 7.8 Infection with HN878, 8 weeks postinfection (H&E). Large granulomas with extensive central necrosis (N), regular as well as foamy histiocytes (H), lymphocytes (L), and Langhans' type multinucleate giant cells (Lg). (a) MP. (b) HP.

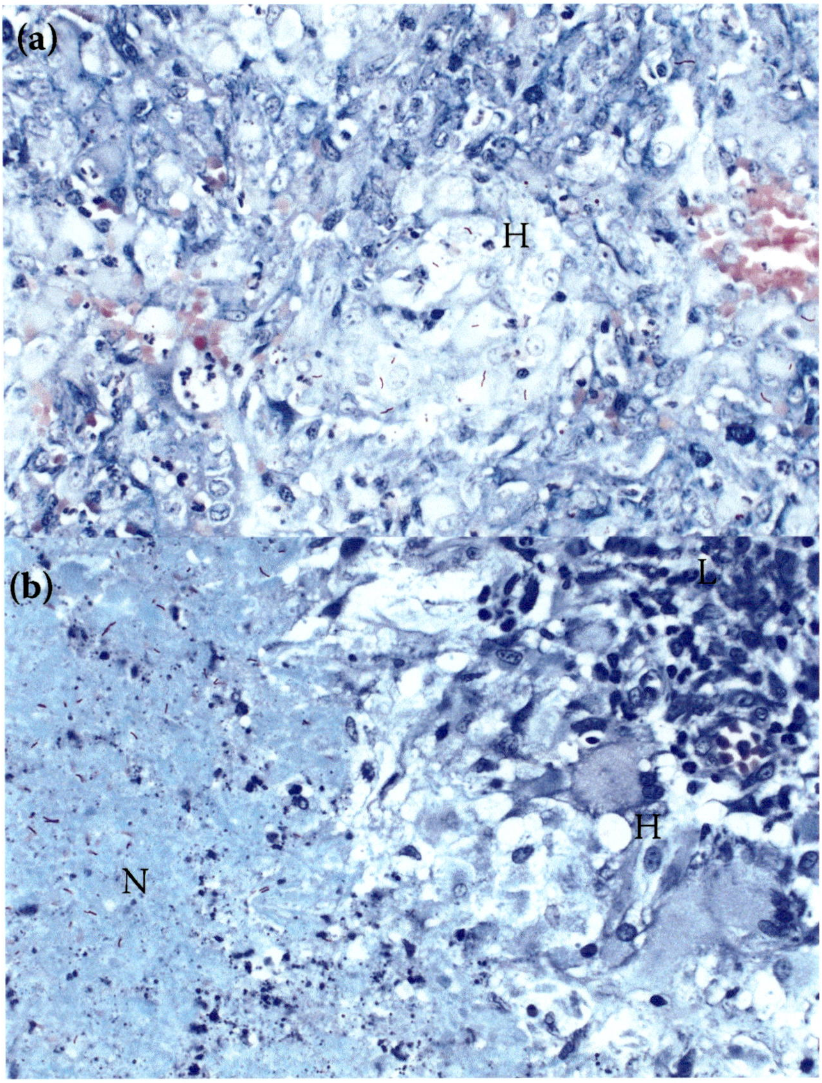

FIGURE 7.9 Infection with HN878 (ZN, HP). Acid-fast bacilli are seen within the histiocytes (H) and in the central necrotic areas (N) of the granulomas. (a) 8 weeks postinfection. (b) 12 weeks postinfection. L—lymphocytes.

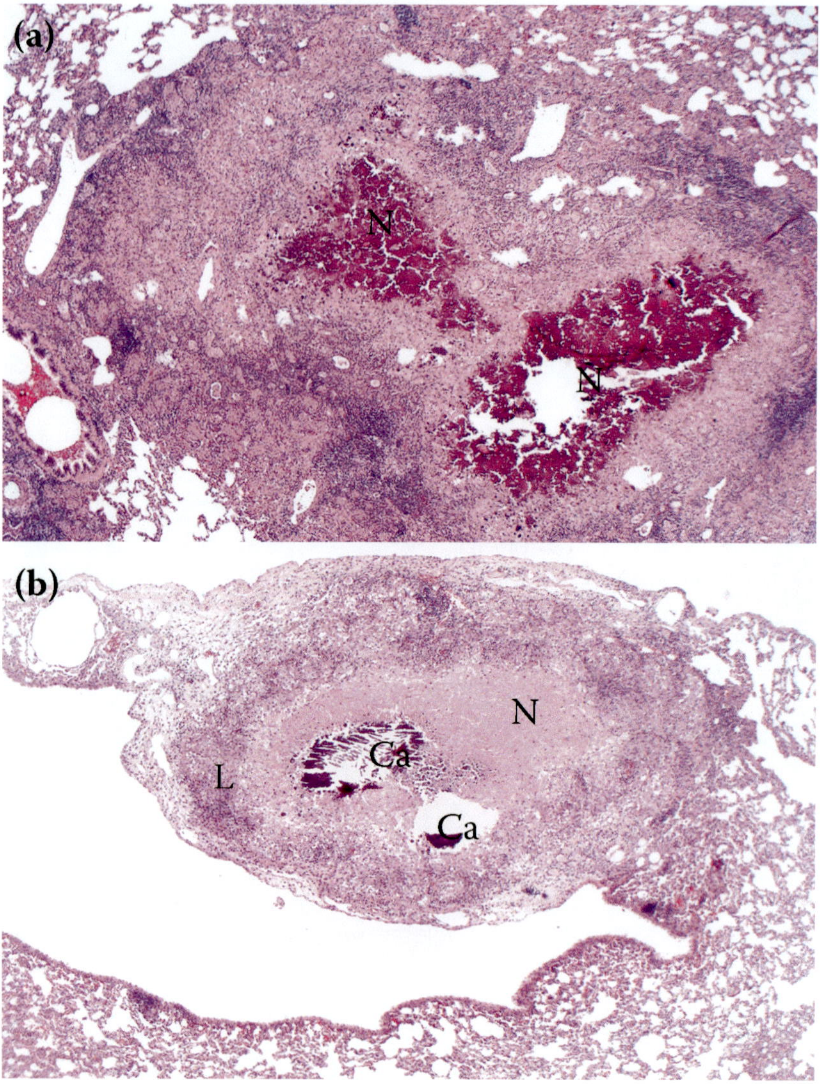

FIGURE 7.10 Infection with HN878 (H&E, LP). (a) MP. 12 weeks postinfection, extensive necrotic areas, partly calcified central areas (N) with surrounding fibrosis. (b) LP. 16 weeks postinfection, calcification (Ca) can be seen. L—lymphocytes and other chronic inflammatory cells. N—necrosis.

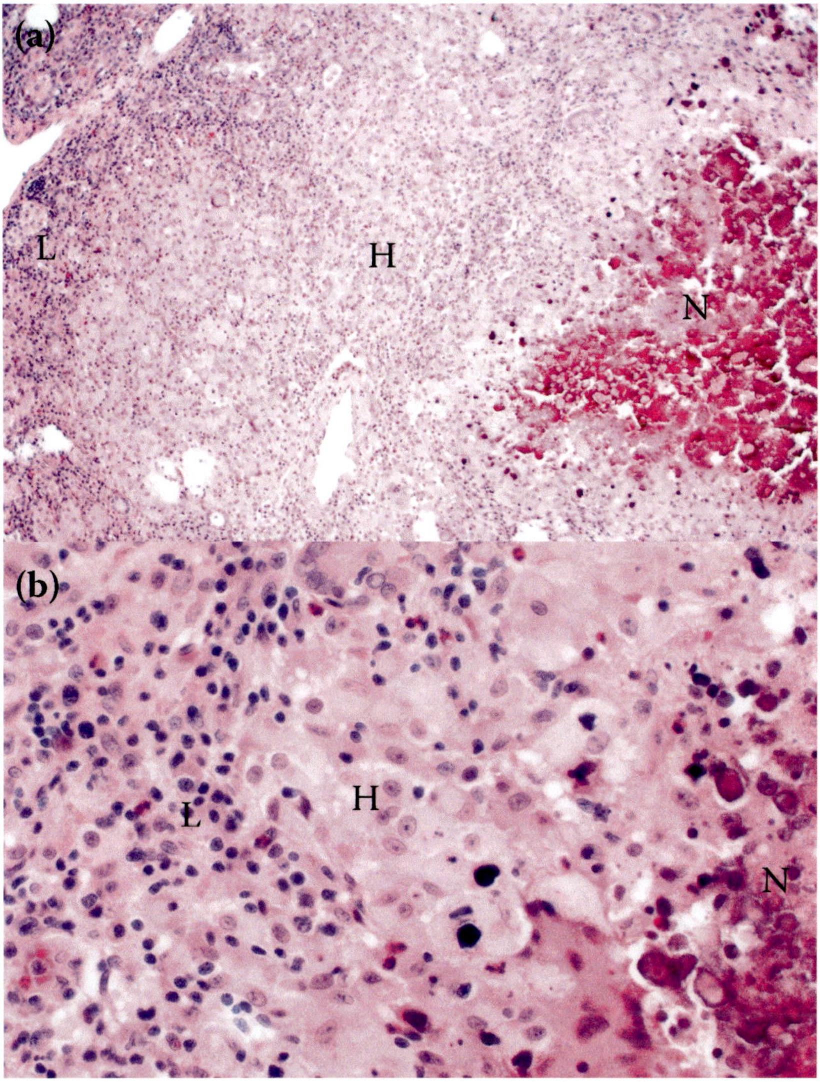

FIGURE 7.11 Infection with HN878, 12 weeks (H&E). (a) MP. (b) HP. Extensive necrosis with calcification (N) is observed. L—mostly lymphocytes with scattered plasma cells. H—epithelioid histiocytes.

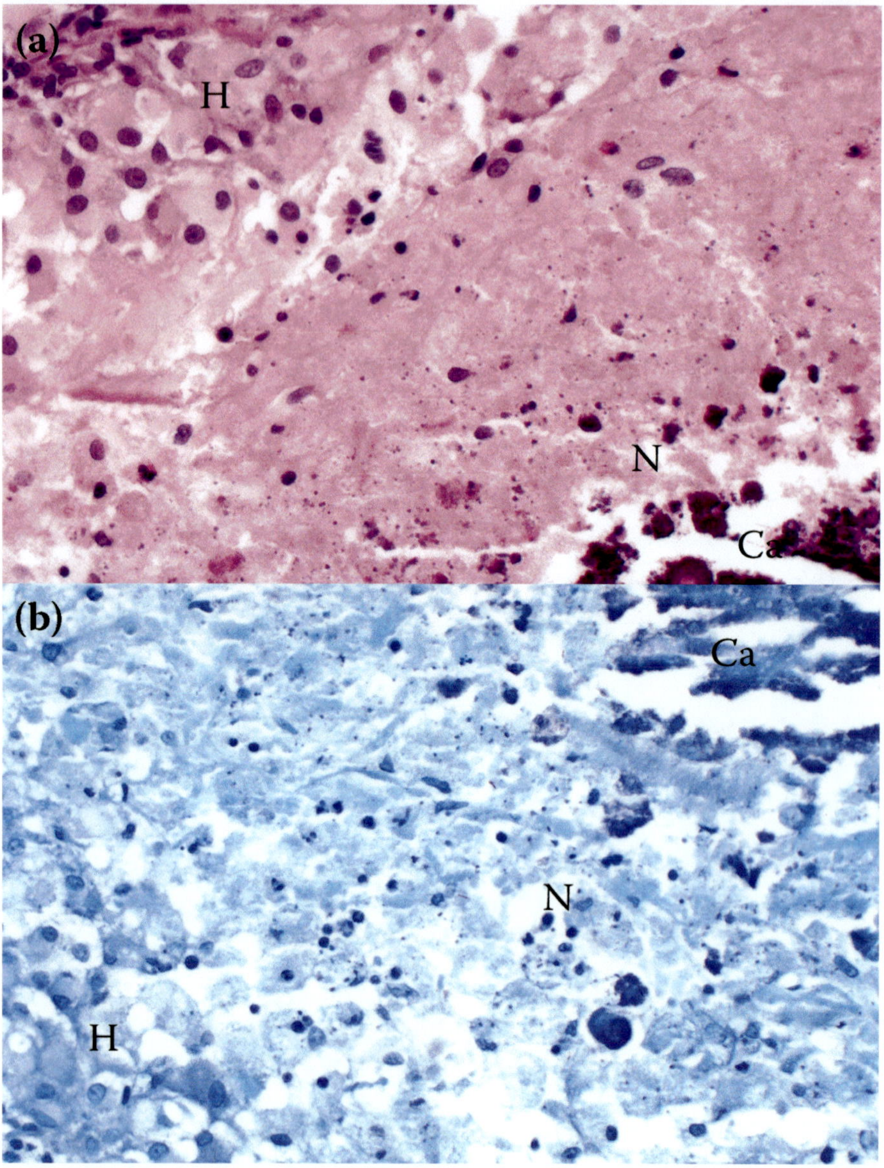

FIGURE 7.12 Calcification. Infection with HN878, 16 weeks. (a) H&E, HP. Mineralization (calcification) can be seen (Ca). (b) ZN, HP. Calcified granulomas had very few acid-fast bacilli (not shown). H—epithelioid histiocytes. N—necrosis. Ca—calcification.

A cross section of the cavities that developed in rabbits infected with HN878 is shown in Figure 7.13. Concentric layers from the cavity out include: an extensive necrotic zone, a thin layer of viable histiocytes mixed in with some dead and damaged cells, a fibrotic area adjacent to a highly cellular granulomatous layer. The granulomatous cellular layer contained large numbers of histiocytes, lymphocytes, and fibroblasts. In the lung periphery between the granulomas, the parenchyma appeared intact but more cellular than normal, with scattered lymphocytes and histiocytes in the alveolar walls and the air spaces. Acid-fast staining revealed numerous acid-fast bacilli (AFB) in the area adjacent to the cavity lumen and in the necrotic zone of the granulomas (Figure 7.14a). Bacilli were usually associated with residual cellular debris and could be seen within some bronchioli (Figure 7.14b). Trichrome stain of the lung sections revealed fibrosis starting at 12 weeks postinfection. Rabbits infected with CDC1551 had significantly less fibrosis (Figure 7.15) compared to the rabbits infected with HN878 (Figure 7.16). These findings were confirmed with the picrosirius red stain, demonstrating significant amounts of collagen around the cavities and extensive neovascularization (Figure 7.17).

Taken together, these results demonstrate that in the New Zealand White outbred rabbit some MTB clinical isolates, such as HN878, give rise to progressive chronic cavitary disease. In contrast, less virulent strains such as CDC1551 are controlled better by the host immune response and are cleared from affected lungs.

STRENGTHS OF THIS ANIMAL MODEL

A major advantage of the rabbit aerosol MTB infection model is that the pathogenesis and disease progression in the lungs of infected animals show important similarities to human disease. First, the rabbit lung effectively models characteristics of MTB latent infection similar to observations made in humans (i.e., after infection with CDC1551, the bacilli are cleared from the lungs and the granulomas disappear). Second, progressive MTB HN878 infection in the rabbit lung is characterized by the development of microbiologic and histopathologic heterogeneous granulomas within the same animal, similar to those described in the human lung. Most importantly, a subset of granulomas undergo suppuration (liquefaction) and give rise to cavities with permissive macrophages supporting the growth of MTB at the luminal surface. Also, the granulomas develop extensive fibrosis and neovascularization. These properties of the rabbit model may facilitate obtaining new insights into the mechanisms of latency and reactivation, as

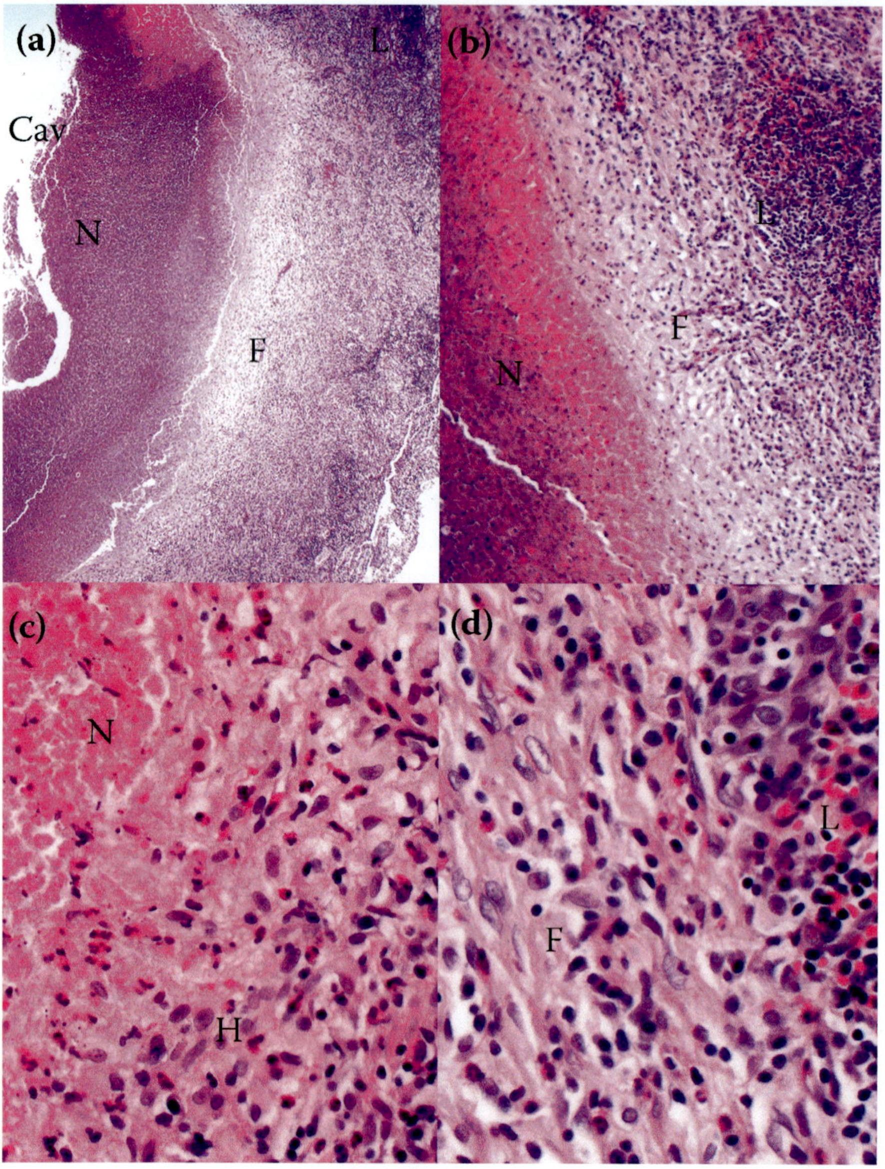

FIGURE 7.13 Infection with HN878, 16 weeks (H&E). (a) Low-power cross section of a cavity (Cav). N—necrosis. F—layer of fibrosis and chronic inflammatory cells. L—lymphocytes. (b) MP. Concentric layers from the inside of the cavity include: an extensive necrotic zone (N), a fibrotic (F) area adjacent to a highly cellular granulomatous layer (L). (c) and (d) HP. Large numbers of histiocytes (H), lymphoplasmacytic infiltrate (L), and fibroblasts (F). N—necrosis.

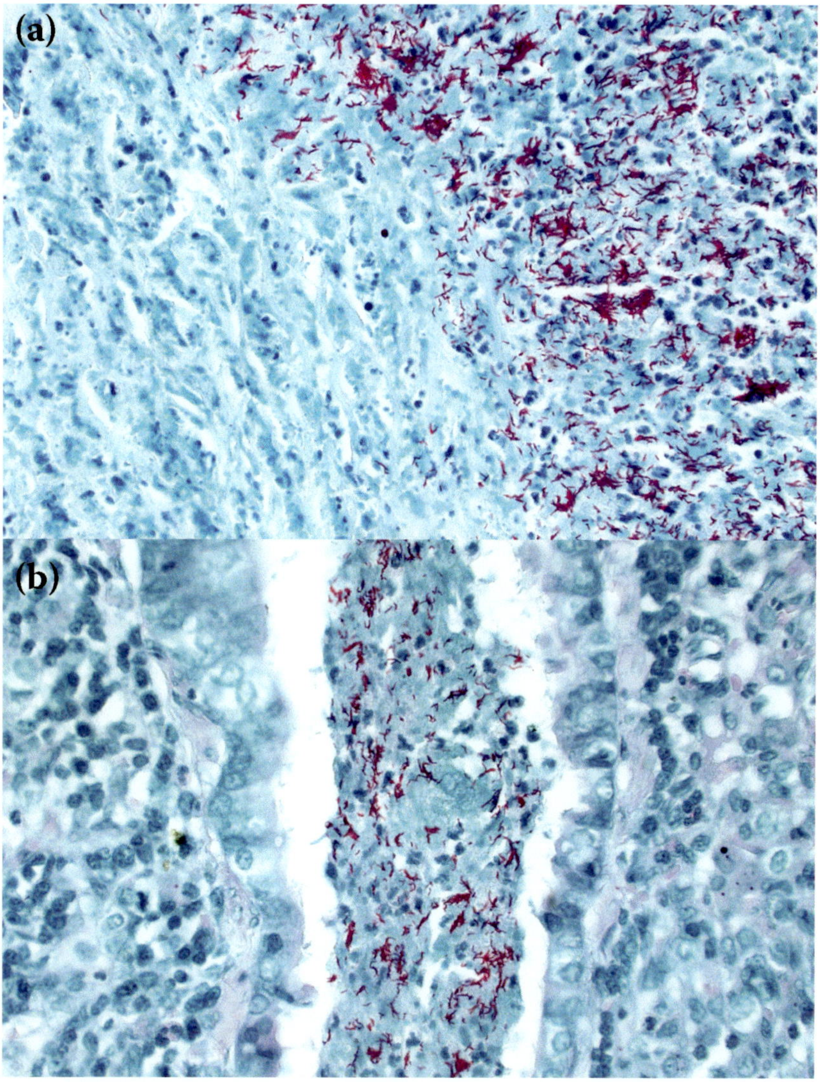

FIGURE 7.14 Acid-fast bacilli. Infection with HN878, 16 weeks (ZN, HP). (a) Large numbers of acid-fast bacilli (seen in magenta) in the area adjacent to the cavity lumen (not shown) and in the necrotic zone of the granuloma. (b) Large numbers of acid-fast bacilli in the cellular debris within a bronchiole.

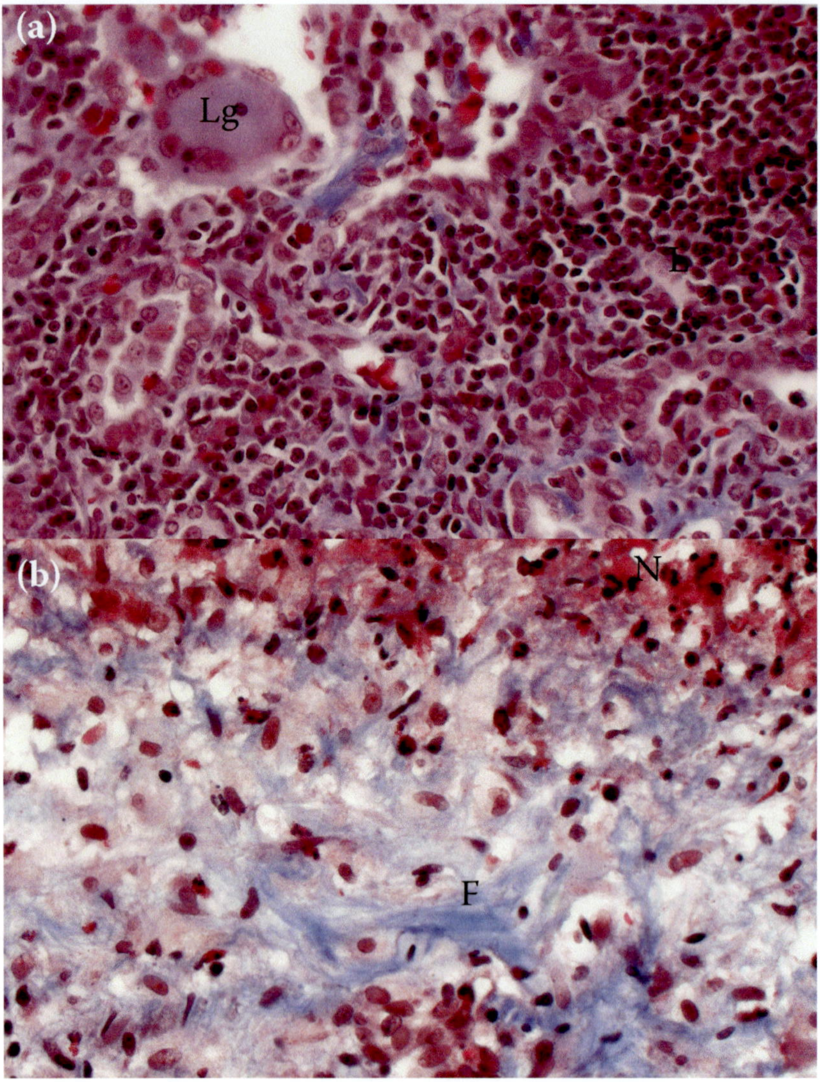

FIGURE 7.15 Infection with CDC1551, 12 weeks (trichrome stain). (a) MP. Lg—Langhans' type giant cell. L—lymphocytes. (b) HP. Early fibrosis (F) is observed. Collagen is stained blue.

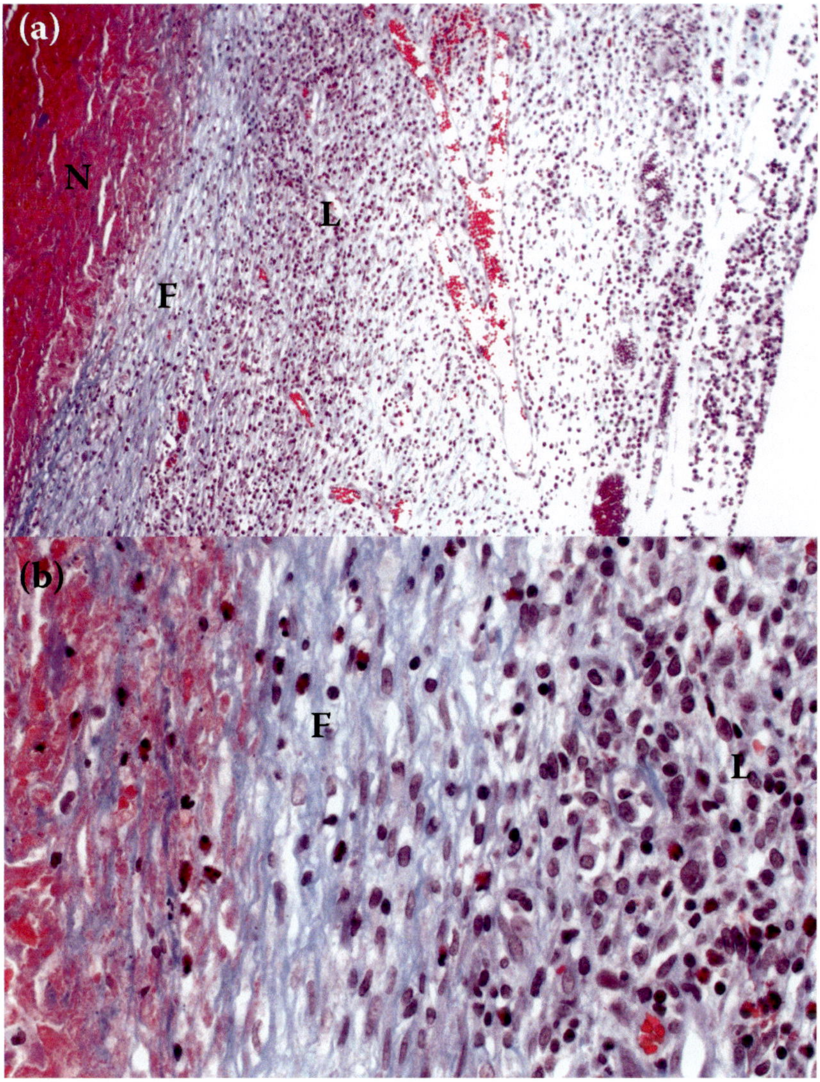

FIGURE 7.16 Wall of a cavitating granuloma. Infection with HN878, 16 weeks (trichrome stain). (a) MP. (b) High-power view demonstrates extensive fibrosis (F) with surrounding lymphocytes and scattered epithelioid histiocytes. N—necrosis. L—lymphocytes.

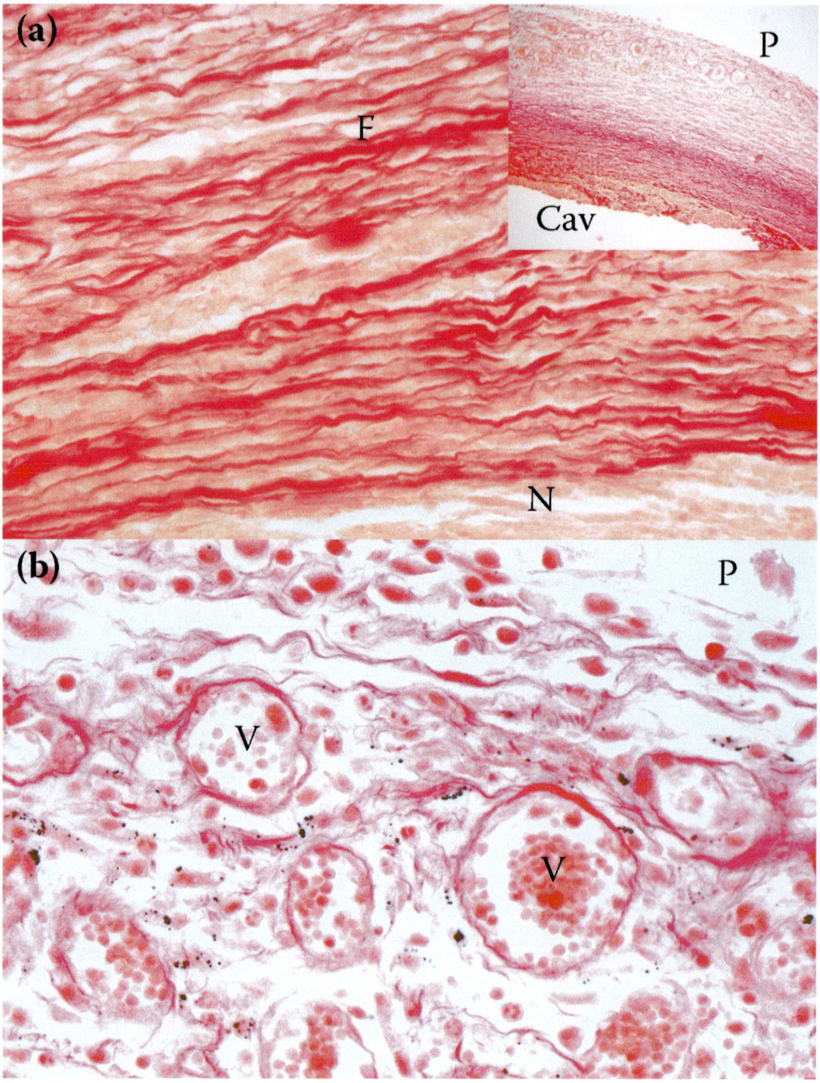

FIGURE 7.17 Wall of a cavitating granuloma. Infection with HN878, 16 weeks (picrosirius stain). (a) and (b) HP, insert—LP. (a) Significant amount of collagen around the cavity (Cav). Collagen is stained in red. (b) Extensive neovascularization. F—fibrosis. N—necrosis. P—pleura. V—vessels.

well as a better understanding of how and why the host immune response fails to control MTB infection, leading to progressive chronic pulmonary disease. The heterogeneity of the lesions facilitates studies on penetration and efficacy of drugs during treatment of latent as well as active disease. Overall, the rabbit model has an important place in the field of tuberculosis research, resembling many aspects of human disease.

SHORTCOMINGS

Currently there is a paucity of immunologic reagents for the rabbit. These reagents are being developed and should enhance the use of this model for the study of tuberculosis.

REFERENCES

Converse, P.J., Dannenberg, A.M., Jr., Shigenaga, T., Mcmurray, D.N., Phalen, S.W., Stanford, J.L., Rook, G.A., Koru-Sengul, T., Abbey, H., Estep, J.E., and Pitt, M.L. 1998. Pulmonary bovine-type tuberculosis in rabbits: Bacillary virulence, inhaled dose effects, tuberculin sensitivity, and *Mycobacterium vaccae* immunotherapy. *Clin Diagn Lab Immunol,* 5, 871–881.

Dannenberg, A.M., Jr. 1994. Rabbit Model of Tuberculosis. In *Tuberculosis: Pathogenesis, protection and control,* ed. B.R. Bloom. Washington, DC: American Society for Microbiology, 149–156.

Dannenberg, A.M., Jr. 2001. Pathogenesis of pulmonary *M. bovis* infection: Basic principles established by the rabbit model. *Tuberculosis,* 81, 87–96.

Dannenberg, A.M., Jr. 2006. *Pathogenesis of human pulmonary tuberculosis: Insights from the rabbit model,* 1st ed. Washington, DC: ASM Press.

Flynn, J.L., Izzo, A., Tsenova, L., and Kaplan, G. 2008. Experimental animal models of tuberculosis. In *Handbook of tuberculosis: Immunology and cell biology,* eds. S.H.E. Kaufmann and W.J. Britton. Weinheim: Wiley-VCH Verlag GmbH & Co. KGaA, 389–426.

Lurie, M.B. 1928. The fate of human and bovine tubercle bacilli in various organs of the rabbit. *J Exp Med,* 48, 155–182.

Lurie, M.B. 1932. The correlation between the histological changes and the fate of living tubercle bacilli in the organs of tuberculosis rabbits. *J Exp Med,* 55, 31–54.

Lurie, M.B. 1964. *Resistance to tuberculosis. Experimental studies in native and acquired defensive mechanisms.* Cambridge: Harvard University Press.

Lurie, M.B., Abramson, S., and Heppleston, A.G. 1952. On the response of genetically resistant and susceptible rabbits to the quantitative inhalation of human type tubercle bacilli and the nature of resistance to tuberculosis. *J Exp Med,* 95, 119–134.

Lurie, M.B., Abramson, S., Swartz, J.B., and Heppleston, A.G. 1950. An evaluation of the method of quantitative airborne infection and its use in the study of the pathogenesis of tuberculosis. *Am Rev Tuberc,* 61, 765–797.

Manabe, Y.C., Dannenberg, A.M., Jr., Tyagi, S.K., Hatem, C.L., Yoder, M., Woolwine, S.C., Zook, B.C., Pitt, M.L., and Bishai, W.R. 2003. Different strains of *Mycobacterium tuberculosis* cause various spectrums of disease in the rabbit model of tuberculosis. *Infect Immun,* 71, 6004–6011.

Manca, C., Tsenova, L., Barry, C.E., 3rd, Bergtold, A., Freeman, S., Haslett, P.A., Musser, J.M., Freedman, V.H., and Kaplan, G. 1999. *Mycobacterium tuberculosis* CDC1551 induces a more vigorous host response in vivo and in vitro, but is not more virulent than other clinical isolates. *J Immunol,* 162, 6740–6746.

Manca, C., Tsenova, L., Bergtold, A., Freeman, S., Tovey, M., Musser, J.M., Barry, C.E., 3rd, Freedman, V.H., and Kaplan, G. 2001. Virulence of a *Mycobacterium tuberculosis* clinical isolate in mice is determined by failure to induce Th1 type immunity and is associated with induction of IFN-alpha/beta. *Proc Natl Acad Sci USA,* 98, 5752–5757.

Medlar, E.M., and Sasano, K.T. 1935. A study of the pathology of experimental pulmonary tuberculosis in the rabbit. *Amer Rev Tuber,* 34, 456–475.

CHAPTER 8

Pulmonary Tuberculosis in the Guinea Pig

Randall J. Basaraba and Ian M. Orme

CONTENTS

INTRODUCTION

Guinea pigs were among the first laboratory animals used as an experimental model of human tuberculosis. Approximately 20 years before Robert Koch used the guinea pig to isolate and ultimately identify the tubercle bacillus as the causative agent of human tuberculosis, John Burton Sanderson and Wilson Fox independently used the guinea pig to document some of the first detailed descriptions of the microscopic lesions of "artificial tuberculosis" (Sanderson, 1867; Fox, 1868; Koch et al., 1982).

In the early 1870s, Edward Emmanuel Klein described studies in which guinea pigs were inoculated with infectious exudate from human tuberculosis patients to study the anatomy of the lymphatic system of the lung and serous membranes (Klein, 1874). Klein documented in detail the lymphatic dissemination and progression of the inflammatory lesions originating from the site of inoculation to other parts of the body. Klein was the first to recognize and document the importance of lymphatic drainage in the spread of *Mycobacterium tuberculosis* (MTB) and described in detail involvement and obstruction of the pulmonary lymphatic vessels by the inflammatory process that is also a prominent feature in guinea pigs infected by low-dose aerosol exposure (Klein, 1873, 1875; Basaraba et al., 2006a,b).

Over the past 50 years the guinea pig model has contributed significantly to our understanding of basic host–pathogen interactions and more recently has been used to evaluate the safety and efficacy of new candidate vaccines and antituberculosis drugs (McMurray, 2001; Horwitz and Harth, 2003; Orme, 2005, 2006; Lenaerts et al., 2007; Basaraba, 2008; Williams et al., 2009). Differences in the lesion morphology and clinical disease outcomes in response to vaccination and drug therapy in guinea pigs has provided important insights into some of the most challenging issues in tuberculosis research, including the mechanisms of resistance conferred by vaccination and the persistence of drug-tolerant bacilli.

Normal Anatomy

The anatomy of the respiratory tract varies among small animal models and can influence how efficiently an infection is established by experimental or natural aerosol exposure. A detailed description of the unique anatomic features and dimensions of the upper airways of the guinea pig have been determined (Schreider, 1983). Some anatomic features of the lung parenchyma of the guinea pig differ from that of humans but have similarities to mice, rats, hamsters, gerbils, and rabbits. Humans have a thick pleural surface, whereas in the guinea pig and other small animal species, the pleural membrane is composed of a thin layer of fibrous connective tissue. In contrast to humans, who have extensive interlobular and segmental connective tissue, guinea pigs have little if any interlobular connective tissue. Similar to humans, guinea pigs have several generations of nonrespiratory (nonalveolarized) bronchioles. In humans, terminal bronchioles end in respiratory bronchioles, whereas in guinea pigs they end in alveolar ducts or short, poorly developed respiratory bronchioles (Figure 8.1). Humans

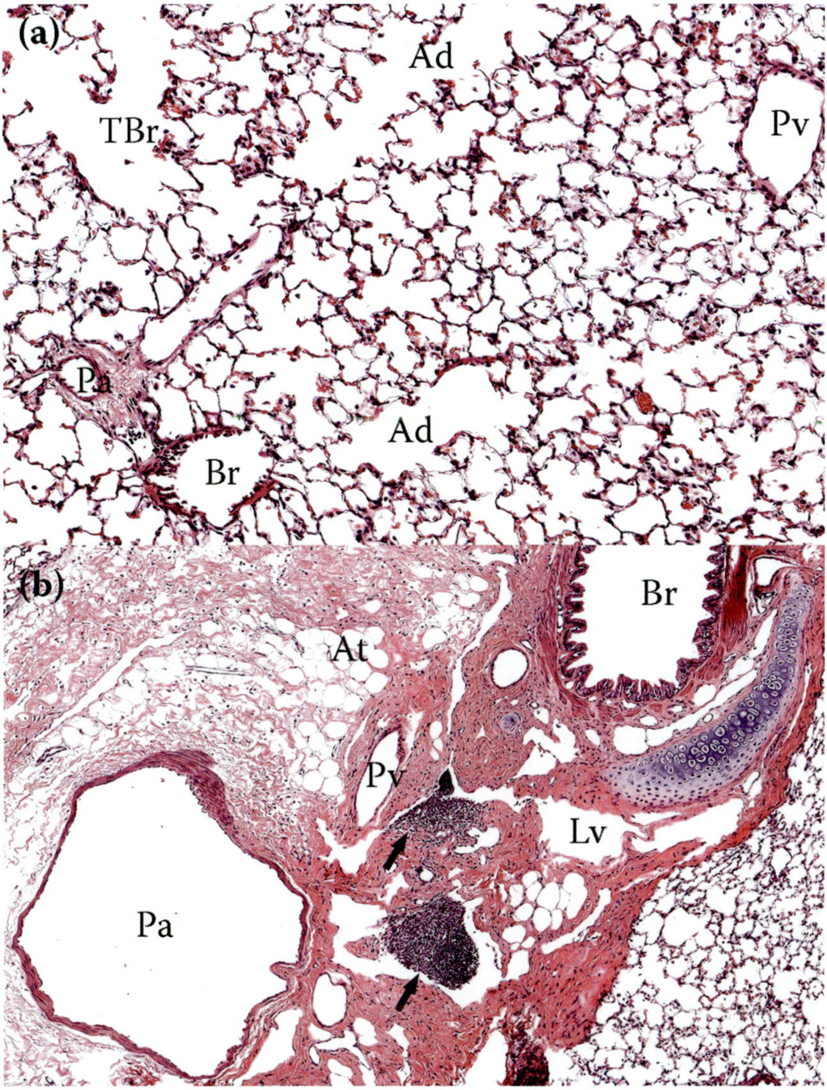

FIGURE 8.1 Normal microscopic anatomy (H&E, LP). (a) Guinea pigs have poorly developed terminal bronchioles (TBr) that end in alveolar ducts (Ad) that serve as respiratory bronchioles. In contrast, nonrespiratory bronchioles (Br) have walls composed of smooth muscle and fibrous connective tissue but are devoid of cartilaginous rings. Both pulmonary veins (Pv) and pulmonary arteries (Pa) are thin walled, with arteries having thicker smooth muscle walls. (b) The dense fibrous connective tissue stroma supports pulmonary arteries (Pa), pulmonary veins (Pv), and bronchi (Br), as well as an extensive network of lymphatic vessels (Lv) that are closely associated with well-developed lymphoid aggregates (arrow). Aggregates of adipose tissue (At) are also found throughout the stroma.

have several generations of alveolarized respiratory bronchioles, whereas in the guinea pig they are absent or shortened (Tyler and Julian, 1992).

Guinea pigs have well-developed lymphoid aggregates that extend throughout the perivascular, peribronchial, and peribronchiolar connective tissue of the lung (Figure 8.2, Figure 8.3). In the guinea pig, lymphoid aggregates adjacent to the large-diameter vessels and airways are supported by fibrous connective tissue and are among the first sites of granuloma development following MTB infection (Klein, 1874, 1875; Basaraba et al., 2006a,b). The lymphoid aggregates are closely associated with an extensive network of lymphatic vessels that directly connect the lung parenchyma to the regional lymph nodes (Figure 8.3).

INOCULUM AND ROUTE OF INOCULATION

TB Strains

Guinea pigs show differences in susceptibility to laboratory and clinical strains of MTB, which is reflected as differences in the rate of clinical disease progression and lesion severity (Balasubramanian et al., 1992a,b; Palanisamy et al., 2008). In the current study, low-dose aerosol exposure of immunologically naïve Dunkin-Hartley guinea pigs to the H37Rv or Erdman KO1 strains MTB was used to establish pulmonary and extrapulmonary infections.

Method of Infection

Low-Dose Aerosol Exposure

It has been estimated that aerosol exposure to as few as 3 to 5 CFU of virulent tubercle bacilli is sufficient to establish a pulmonary and extrapulmonary infection in guinea pigs (Smith et al., 1991). The Madison aerosol generating device (Generation III) was originally designed in the 1970s specifically for guinea pig aerosol infections and is still in use today (Smith and Harding, 1977). The device consists of a stainless steel chamber that supports a basket insert of 18 separate compartments designed to hold individual animals. The low-dose infection protocol is designed to deliver approximately 30 to 50 bacilli during a 15-minute exposure by nebulization.

FEATURES

The clinical and pathologic manifestation of MTB infection in guinea pigs is dependent on a wide variety of host and pathogen factors. The immune status of the host as well as the virulence, dose, and origin of the challenge

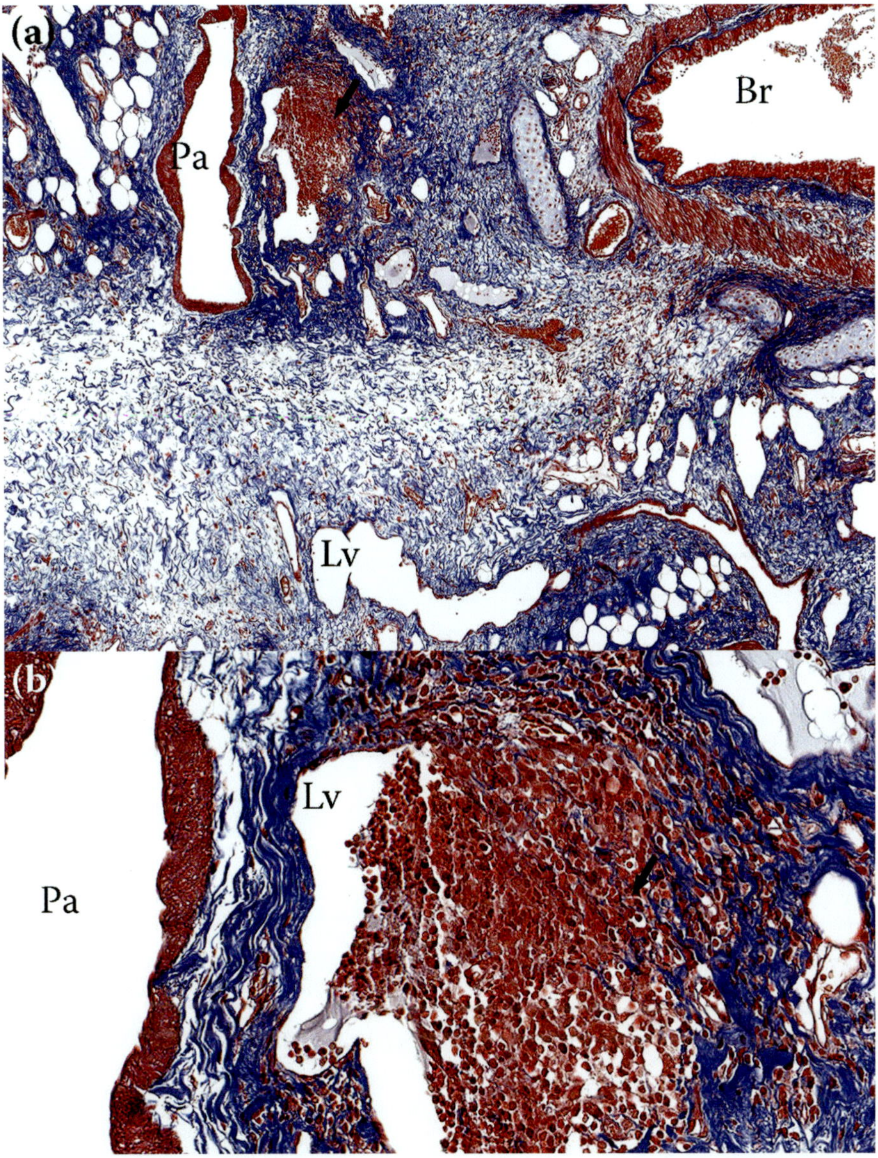

FIGURE 8.2 Normal anatomy of lymphoid aggregates in the supporting connective tissue of the guinea pig lung (Masson's trichrome stain, MP, and HP). (a) MP. Extensive fibrous connective tissue stroma supports pulmonary arteries (Pa), bronchi (Br), and extensive network of lymphatic vessels (Lv) and well-developed lymphoid aggregates (arrow). (b) HP. Lymphoid aggregates (arrow) are closely associated with lymphatic vessels (Lv), all of which are supported by dense fibrous connective tissue (blue). They correspond to bronchus-associated lymphoid tissue in human lung anatomy.

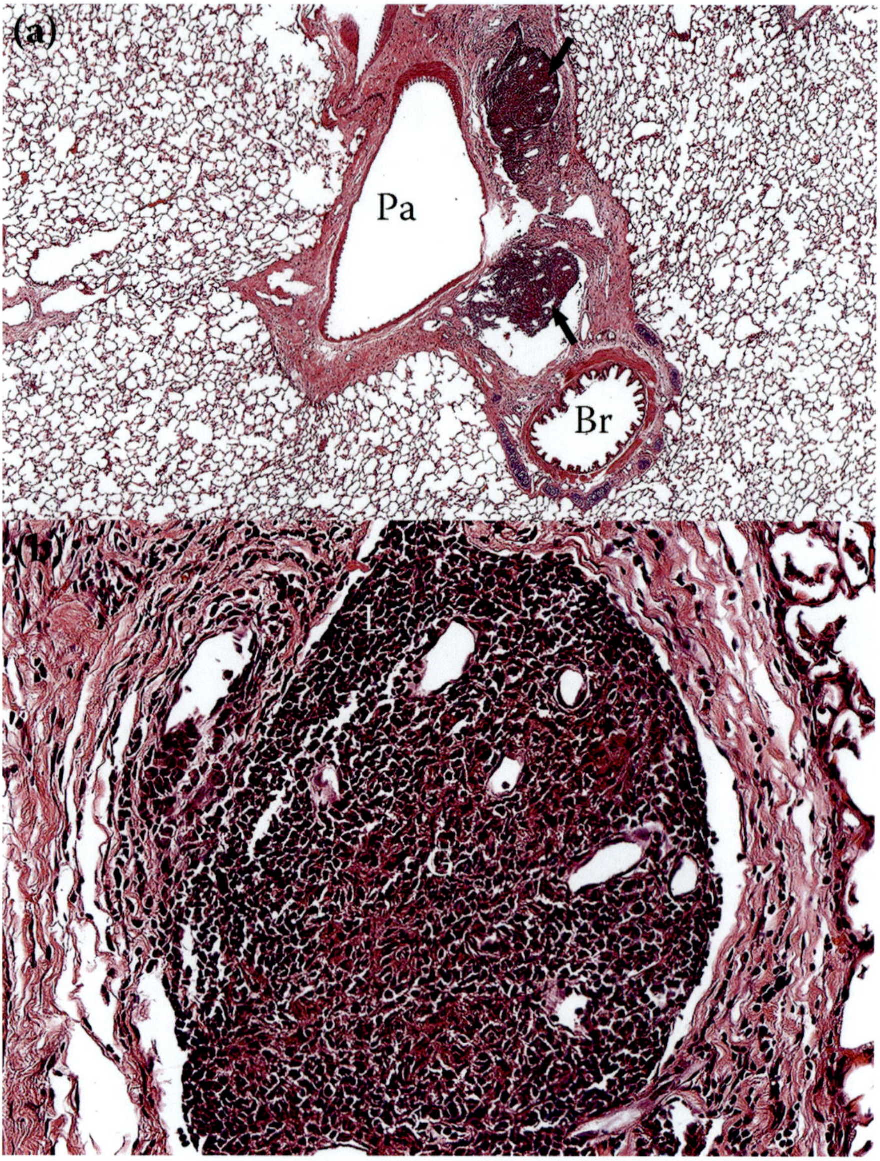

FIGURE 8.3 Granuloma formation in lymphoid aggregates in the early stages of MTB infection in the guinea pig (H&E, LP, and HP). (a) LP. The lymphoid aggregates (arrows) within the connective tissue stroma that supports pulmonary arteries (Pa) and bronchi (Br) are among the first sites of primary lesion development following aerosol infection of guinea pigs with MTB. (b). HP. Sheets of lymphocytes (L) are interrupted by early infiltrates of histiocytes that form the initial granuloma (G) by day 20 of infection.

strain have a significant influence on the progression of clinical disease and lesion morphology (Smith and Wiegeshaus, 1989; McMurray, 2003; Basaraba et al., 2006a; Ordway et al., 2008).

Acute Infection

Following low-dose aerosol exposure of guinea pigs with MTB, the initial sites of inflammation (primary lesions) develop in at least three distinct anatomic locations throughout the lung. They can be basically classified as subpleural, deep parenchymal, or those associated with the perivascular or peribronchial lymphoid aggregates and pulmonary lymphatic vessels. Lesions are rarely visible by day 5 of infection, but by day 15 histiocytes, granulocytes, and lymphocytes form poorly organized alveolar and interstitial infiltrates, often with fibrin exudation (Basaraba et al., 2006b). Mononuclear histiocytes become the prominent cell type during early granuloma formation between days 15 and 20. Similar cells infiltrate perivascular or peribronchial lymphoid aggregates without altering the normal structures initially, but eventually progress to efface normal follicular architecture and expand to involve the adjacent pulmonary lymphatic vessels themselves (lymphangitis) by day 20 of infection (Figure 8.3) (Klein, 1875; Basaraba et al., 2006b). Deep parenchymal and subpleural inflammatory foci eventually organize into distinct granulomas and show the first evidence of central necrosis between days 20 and 30 (Figure 8.4). Central lesion necrosis is a prominent feature of primary granulomas that are composed mostly of histiocytes with fewer lymphocytes and granulocytes by day 30, a process that peaks between 30 and 60 days of infection (Figure 8.5). Even in the early stages of granuloma formation, the loss of pulmonary parenchymal architecture is accompanied by fibrous encapsulation and collagen deposition (Figure 8.6). By day 20 of infection, early granulomas, similar to those seen in the lung, form in the cortex of draining tracheobronchial and anterior mediastinal lymph nodes as part of the primary lesion complex (Basaraba et al., 2006a). By day 30 of infection, early secondary lesions (tuberculous pneumonia) develop following reinfection of the lung as a consequence of hematogenous dissemination of bacilli (Balasubramanian et al., 1994; McMurray, 2003; Ordway et al., 2007).

Chronic Infection

Between 30 and 60 days of infection the expansion of primary lesions slows, but during this bacillemic phase of the infection, the remainder of the

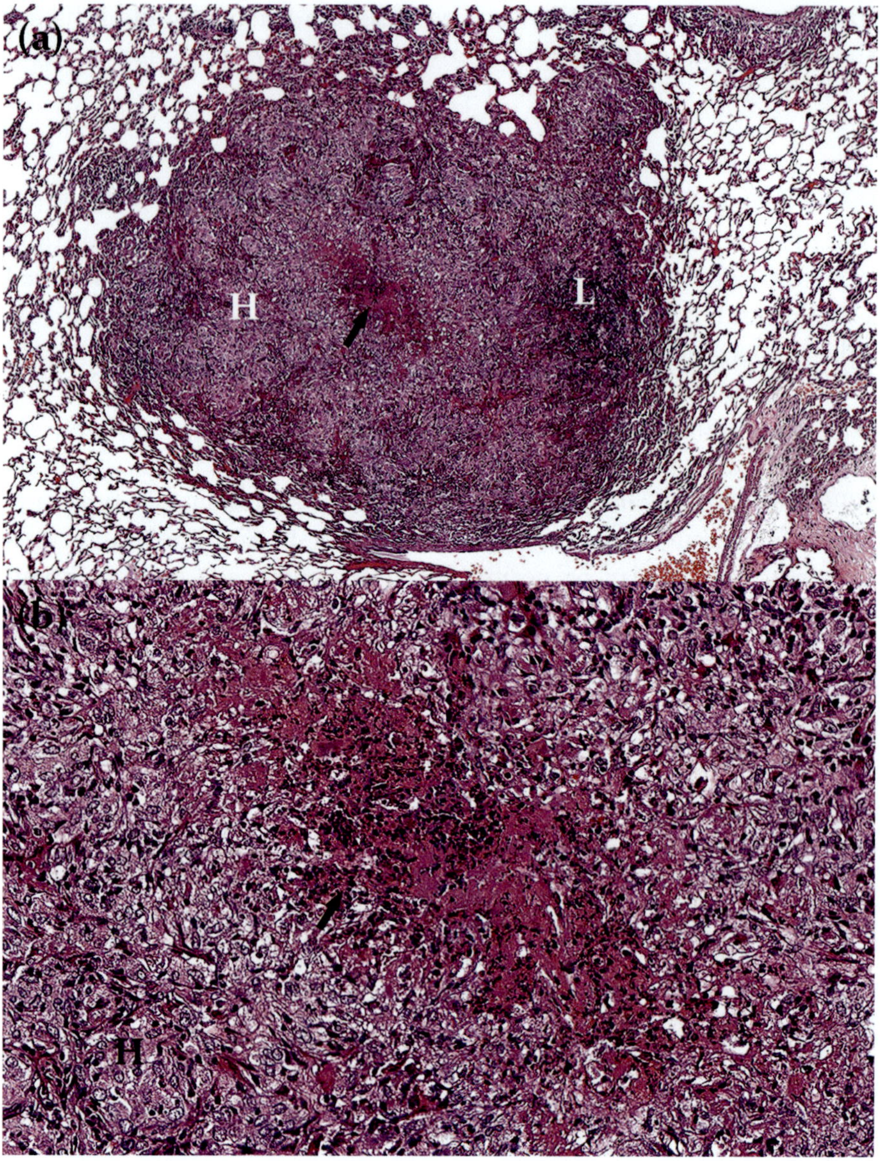

FIGURE 8.4 Primary lesions in immunologically naïve guinea pigs develop central lesion necrosis (H&E, LP, and HP). (a) LP. Between days 20 and 30 of MTB infection of the guinea pig, primary lesions are composed primarily of histiocytes (H) and fewer lymphocytes (L), which are the future site of infarct-like necrosis (arrow). (b) HP. The early stage necrosis (arrow) is characterized by a loss of histiocyte detail, which is replaced by eosinophilic cytoplasmic debris and basophilic nuclear fragments.

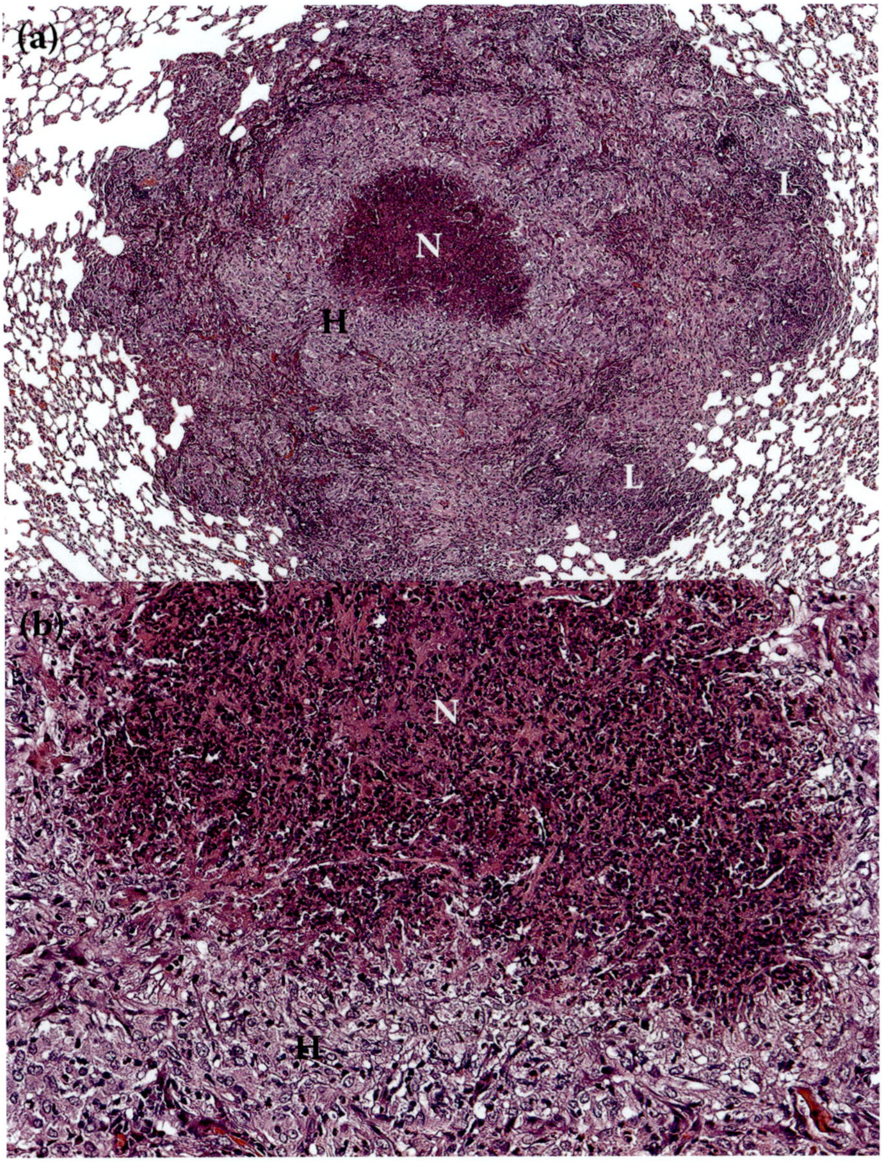

FIGURE 8.5 Central necrosis of primary lesions compresses the surrounding alveoli as lesions expand within the parenchyma over time (H&E, LP, and HP). (a) Between days 30 and 60 of MTB infection of the guinea pig, primary lesion necrosis (N) is composed primarily of histiocytes (H) and fewer peripherally located lymphocytes (L). (b) In advanced necrosis (N) the loss of cytoplasmic detail is extensive and is accompanied by nuclear fragmentation and early calcification.

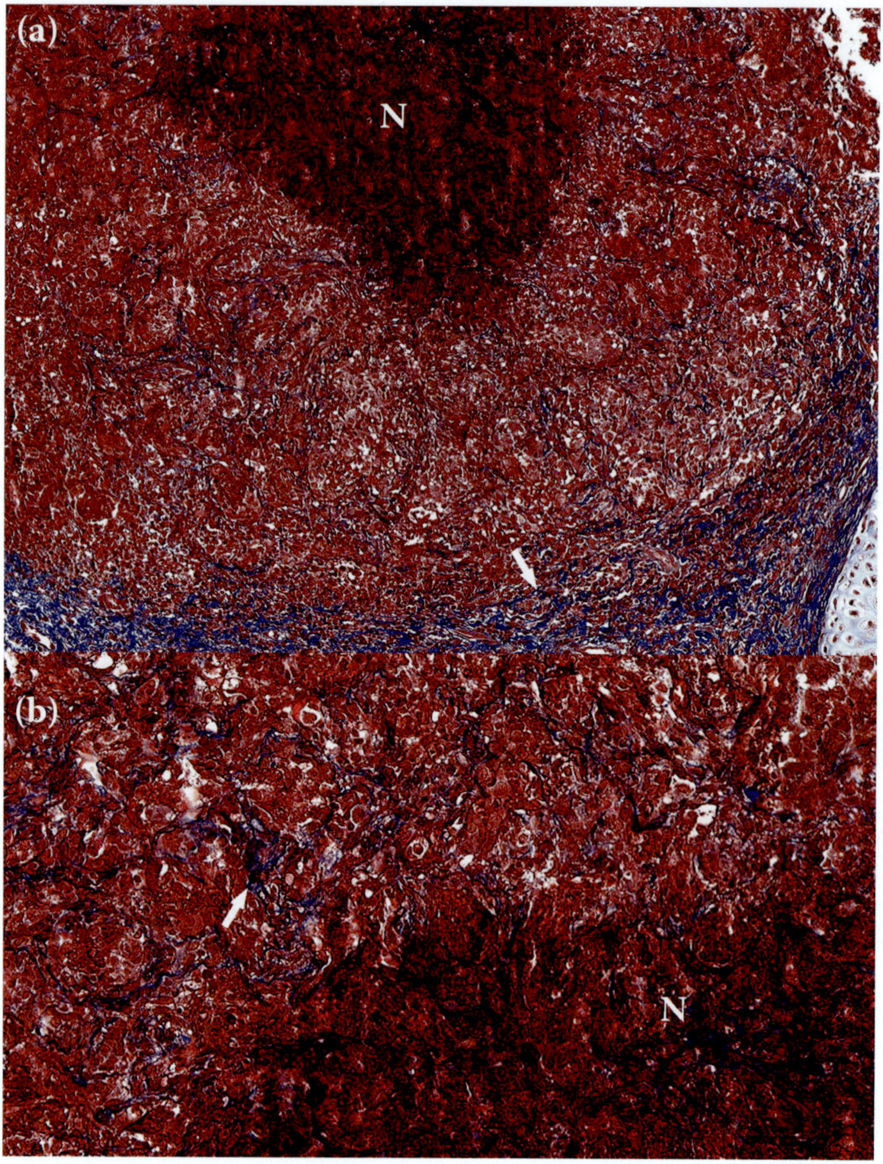

FIGURE 8.6 Fibrous connective tissues develop early as primary lesions expand (Masson's trichrome). (a) MP. Pseudo-encapsulation of primary lesions with necrosis (N) can result from preexisting fibrous connective tissue (arrow) or develop with the developing lesion. (b) HP. Strands of fibrous connective tissue (blue staining, arrow) can be seen throughout the lesion including the center that is undergoing necrosis (N).

parenchyma becomes infiltrated by histiocytes, lymphocytes, and fewer granulocytes with the formation of tuberculous pneumonia (Figure 8.7). Tuberculous pneumonia differs morphologically from primary lesions in that they lack necrosis until the very late stages of infection (Figure 8.8). The progressive pneumonia eventually replaces the pulmonary parenchyma and is accompanied by extensive fibrosis, epithelial regeneration, and a pattern of coalescing lesion necrosis as a consequence of generalized disruption of the pulmonary blood supply (Figure 8.8). In the chronic stages of disease, the original primary lesions of the subpleural and deep parenchyma (Figure 8.9), and less often the perivascular and periperibronchial connective tissue (Figure 8.10), partially heal by dystrophic calcification. Calcification of necrotic primary lesions is one feature that differentiates late-stage necrosis of tuberculous pneumonia, which usually lacks significant calcification (Figure 8.8). At this stage of infection acid fast–positive bacilli are rarely seen in histiocytes associated with tuberculous pneumonia, but are more often concentrated in degenerate macrophages and neutrophils within airway lumens and extracellularly in primary lesions with necrosis (Figure 8.11) (Lenaerts et al., 2007; Basaraba, 2008). Calcification of primary lesions is often incomplete, leaving a rim of residual necrosis that demarcates the calcified center from viable histocytes of the granuloma inflammatory zone.

The importance of primary lesions with necrosis in the guinea pig model is that they represent foci of irreversible tissue damage, which creates a unique microenvironment that harbors persistent bacilli even following combination drug therapy (Lenaerts et al., 2007). Six weeks of drug therapy in guinea pigs is effective at reversing and preventing the development of tuberculous pneumonia. However, if drug therapy is initiated after primary lesions become established, foci of necrosis fail to resolve and partially calcify, leaving an uncalcified rim of residual necrosis (Figure 8.12). These lesions harbor persistent, extracellular bacilli that lead to disease reactivation when antituberculous drugs are discontinued.

Extrapulmonary Lesions

Another important feature of chronic MTB infections in guinea pigs is widely disseminated extrapulmonary lesions, the severity of which is also dependent on the immune status of the host and the virulence of the challenge strain (Palanisamy et al., 2008). The earliest and most extensive extrapulmonary lesions develop in the draining lymph nodes following afferent lymphatic drainage from the lung followed by hematogenous

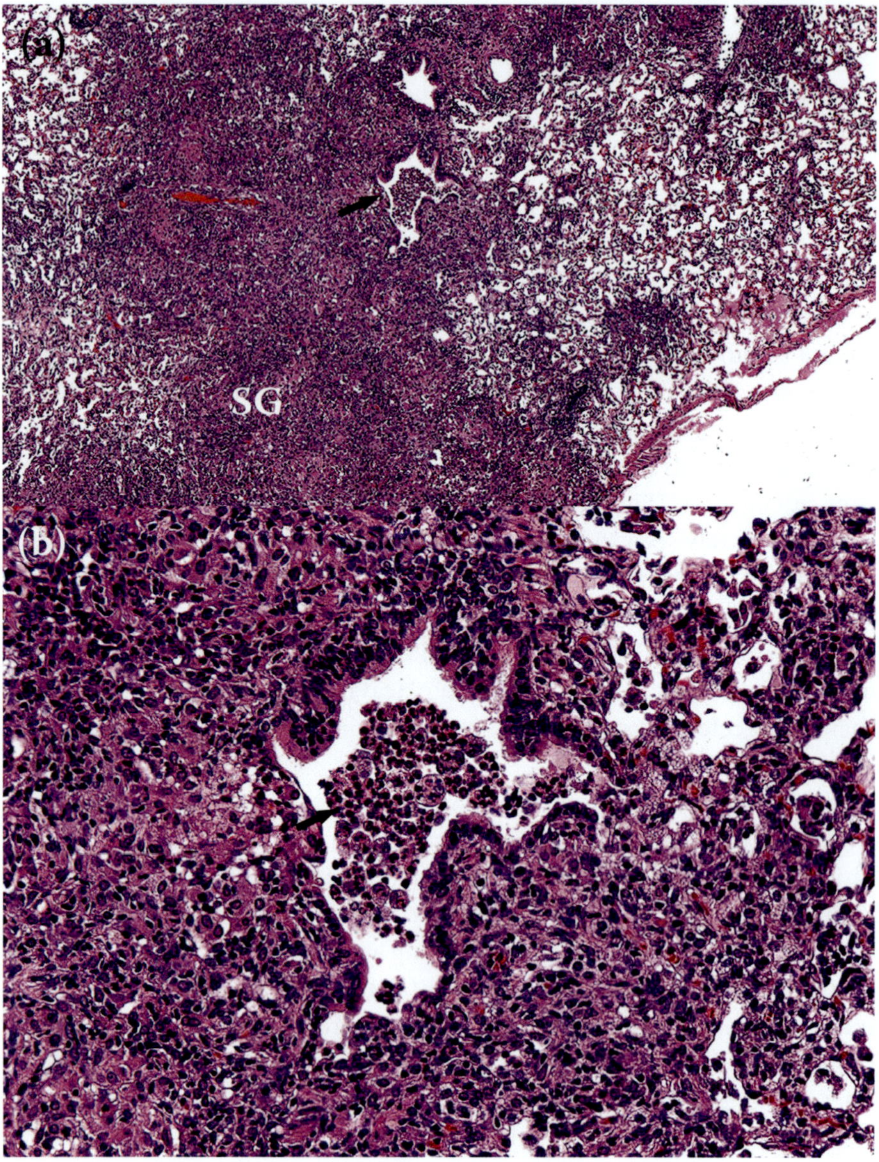

FIGURE 8.7 Secondary granulomatous inflammation is morphologically distinct from primary granulomas (H&E, LP, and HP). (a) Secondary granulomatous inflammation (SG) differs from primary granulomas by being a mixture of histiocytes and lymphocytes that coalesce and infiltrate the normal parenchyma with indistinct lesion margins. Another feature of secondary granulomatous inflammation is the filling of small airways with inflammatory exudate (arrow). (b) A higher magnification of secondary granulomatous inflammation showing the airway exudates (arrow) consists of degenerate neutrophils and macrophages.

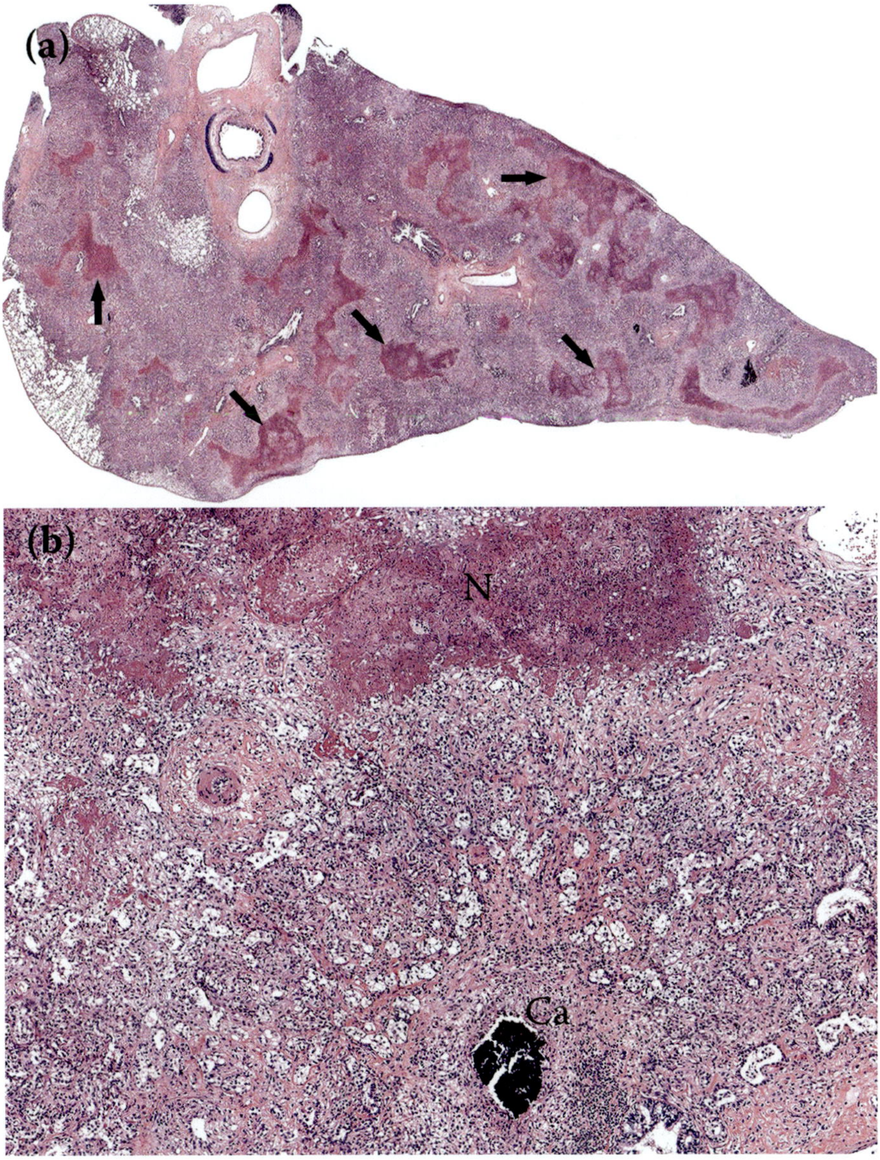

FIGURE 8.8 Secondary granulomatous inflammation. During the chronic stages of disease, lesions develop necrosis that is distinct from primary lesions (H&E, LP, and MP). (a) In a section of lung with greater than 90% involvement, secondary granulomatous inflammation has multiple foci of noncalcified necrosis (arrows) that coalesce with indistinct margins. (b) A higher magnification of secondary granulomatous inflammation with necrosis (N). Secondary lesion necrosis (N) differs from residual primary lesion necrosis that more readily heals by dystrophic calcification (Ca).

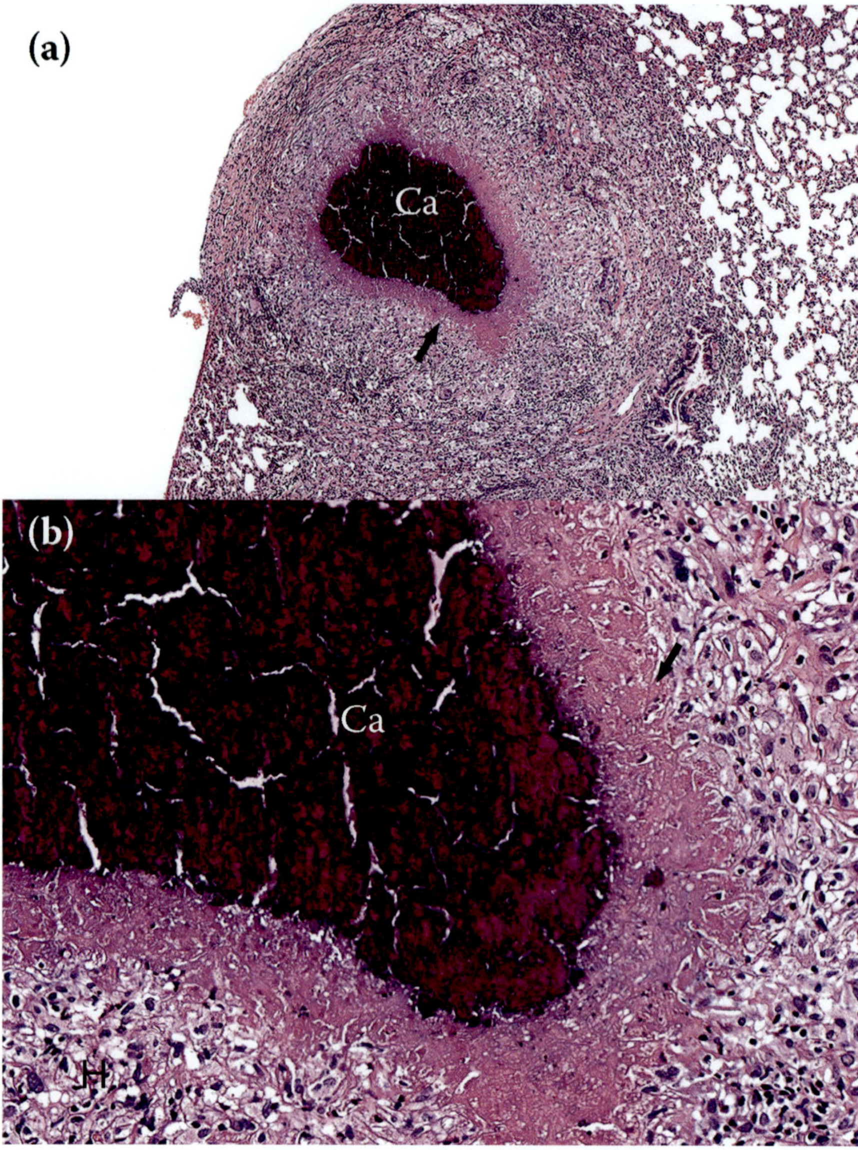

FIGURE 8.9 The necrotic focus of primary lesions partially heals with deposits of a calcified material (H&E, LP, and HP). (a) LP. A subpleural primary lesion with central calcification (Ca). A rim of uncalcified residual necrosis (arrow) demarcates dystrophic calcification from the inflammatory zone. (b) HP. A higher magnification of the calcified lesion shows a rim of noncalcified residual necrosis (arrow) that is sharply demarcated from viable histiocytes (H) within the granuloma.

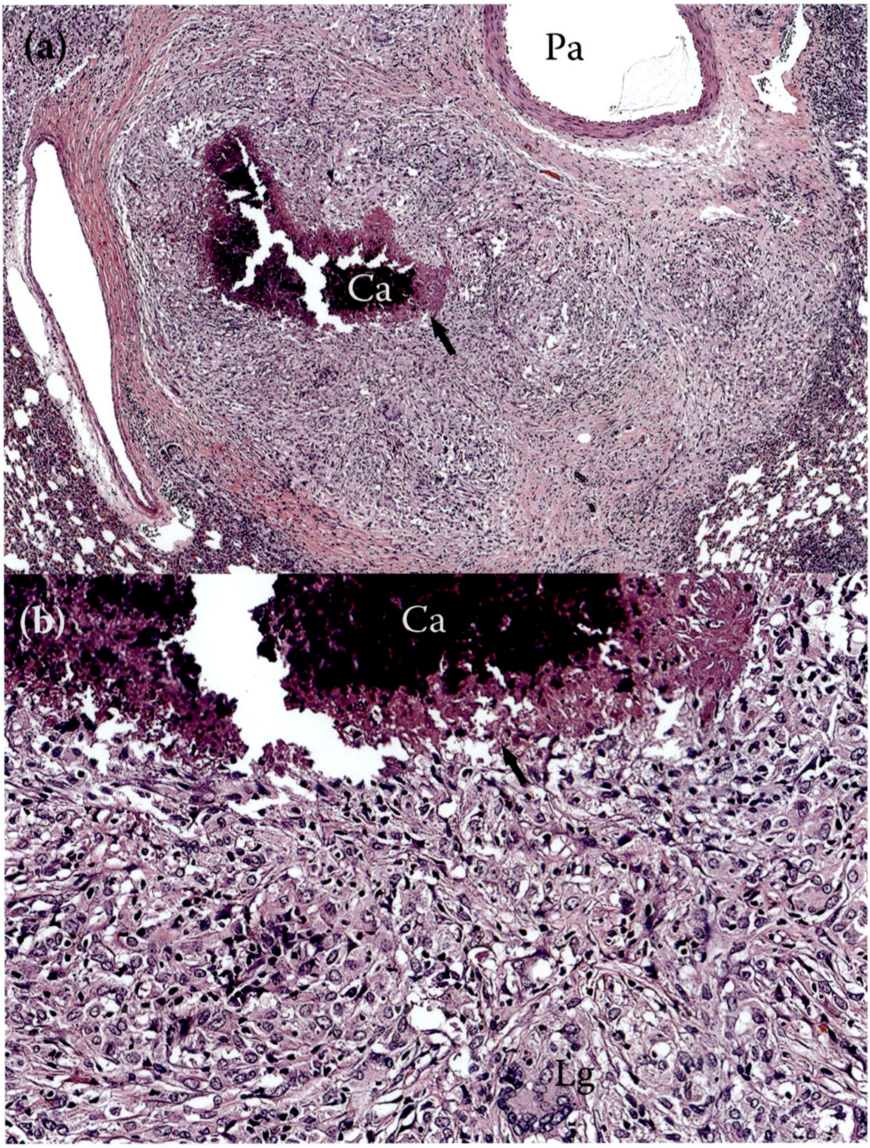

FIGURE 8.10 Advanced granuloma formation with necrosis and calcification in the perivascular connective tissue (H&E, LP, and HP). (a) Adjacent to pulmonary arteries (Pa) and large airways, granulomas become extensive and can progress to central necrosis (arrow) with calcification (Ca). (b) Residual lesion necrosis (arrow) demarcates the calcified center (Ca) from the active histiocytes that comprise the granuloma. Mononuclear histiocytes predominate but are occasionally admixed with multinucleated Langhans' type multinucleate giant cells.

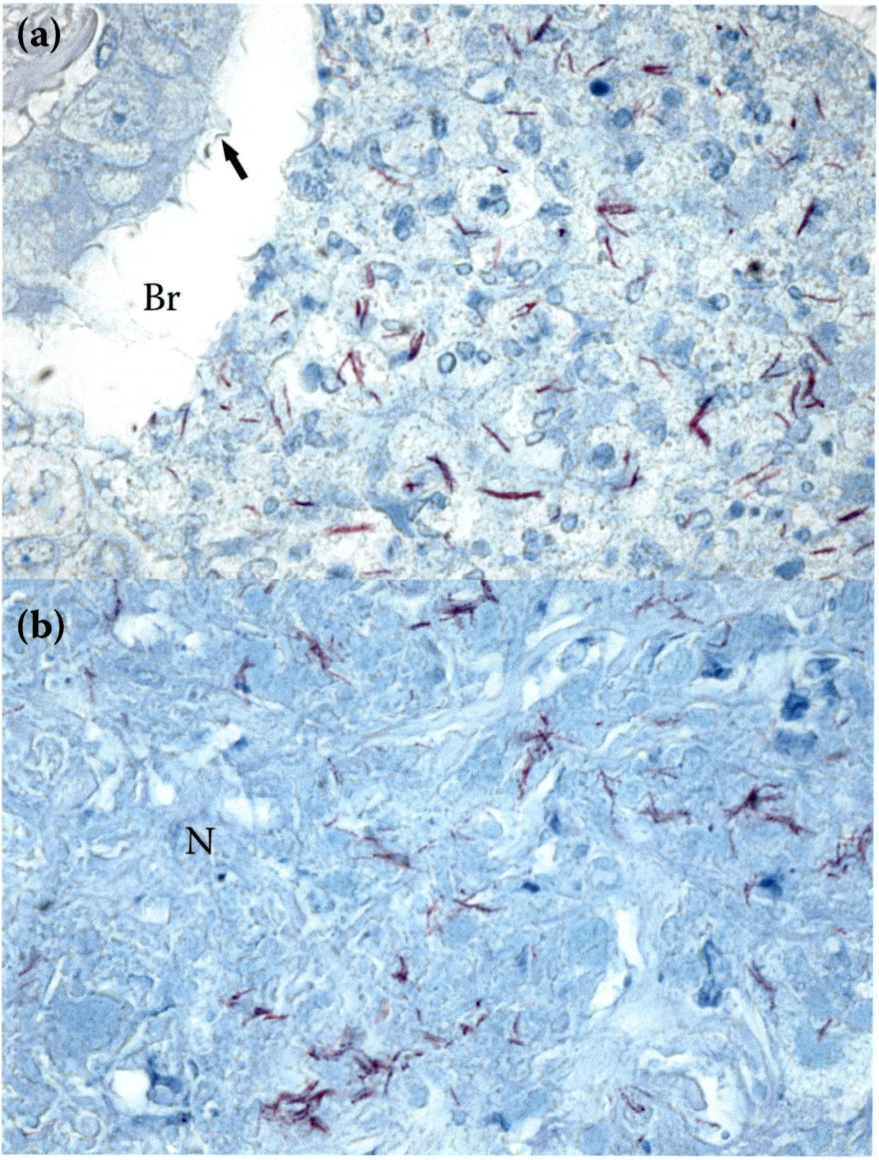

FIGURE 8.11 Acid-fast bacilli are concentrated in airway exudates and extracellular within necrotic foci (ZN oil, HP). (a) Within respiratory epithelial (arrow) lined bronchi (Br), acid-fast bacilli are both intracellular in histiocytes as well as extracellular in airway exudates composed of degenerate macrophages and neutrophils. (b) Within foci of primary lesion necrosis (N), clusters of acid-fast bacilli are extracellular admixed with cytoplasmic and nuclear debris from necrotic inflammatory cells.

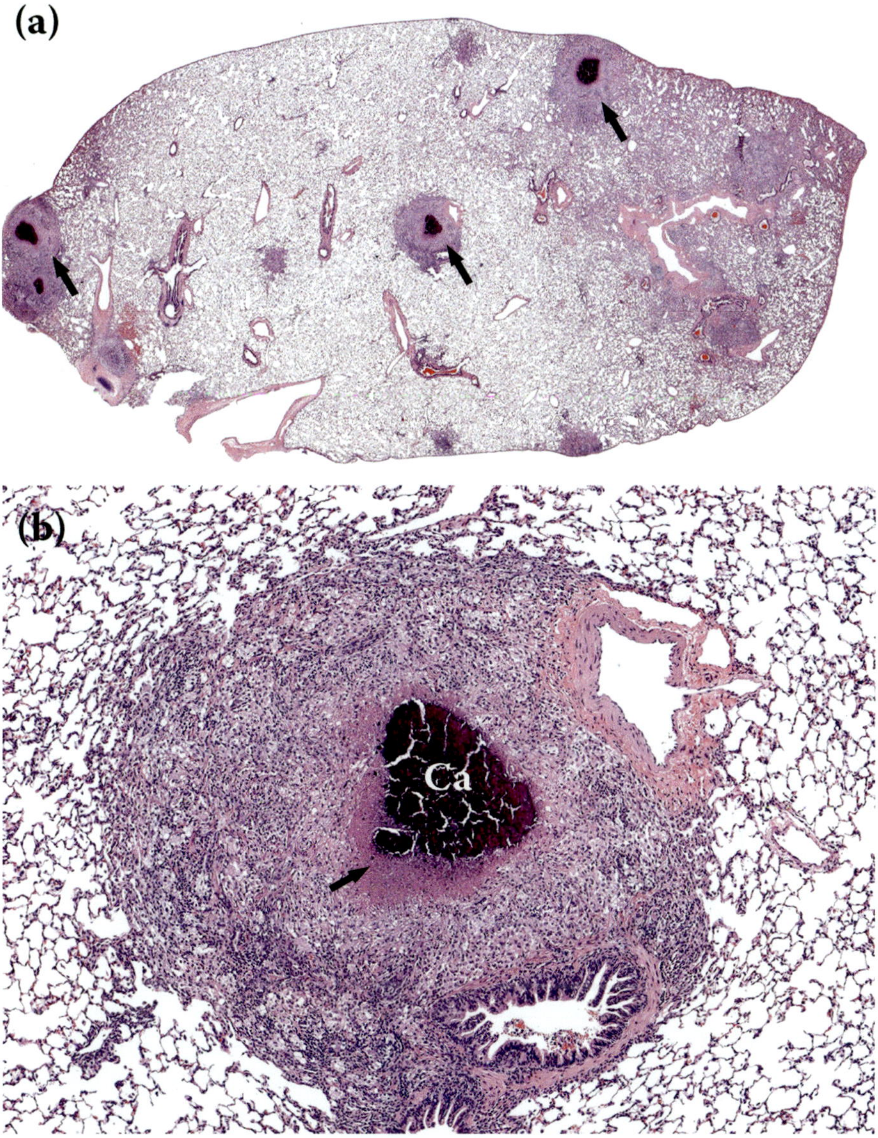

FIGURE 8.12 Drug treatment of MTB-infected guinea pigs is ineffective at clearing primary lesions with necrosis (H&E, LP, and MP). (a) LP. Initiation of combination drug therapy (INH/RIF/PZA) 4 weeks after infection prevents the development of secondary granulomatous inflammation but leaves primary lesions (arrows) unaffected even after 6 weeks of therapy. (b) MP. A higher magnification of primary lesions with dystrophic calcification (Ca) and uncalcified residual necrosis (arrow).

dissemination to other extrapulmonary sites during the bacillemic phase of infection (McMurray, 2003; Basaraba et al., 2006a). The lymph nodes that drain the lung are scattered throughout the peribronchial and peritracheal loose connective tissue of the anterior mediastinum (Figure 8.13). Involvement of draining lymph nodes occurs early in the infection and constitutes part of the primary lesion complex following aerosol infection. Focal to multifocal granulomatous inflammation of the lymph nodes (lymphadenitis) progresses rapidly to replace the normal architecture and becomes necrotic, occasionally with cavitation, between 30 and 60 days of infection (Figure 8.14).

The spleen is among the first extrapulmonary organs that become infected during the bacillemic phase of infection. Lymphocytes in the normal spleen are concentrated in the periarteriolar lymphatic sheaths (Figure 8.15) and are the sites of initial granuloma formation following hematogenous dissemination of bacilli. In the chronic stages of extrapulmonary disease, the majority of periarteriolar lymphatic sheaths are effaced by granulomatous inflammation that ultimately progresses to necrosis and occasionally dystrophic calcification (Figure 8.16).

Latent Disease

Because of the poor natural resistance of guinea pigs to even low-dose MTB exposure, latent disease cannot be established unless the infection is controlled for a period of time with drug therapy. Extracellular bacilli persist in primary lesions with necrosis and serve as the site of reactivation disease when drug therapy is discontinued (Lenaerts et al., 2007).

STRENGTHS OF THIS ANIMAL MODEL

Guinea pigs are among the few small animal tuberculosis models that develop necrotic lesions that harbor persistent, extracellular bacilli even following combination drug therapy. The guinea pig is an excellent small animal model to study the pathogenesis of primary lesion necrosis, extrapulmonary dissemination, and the mechanisms of phenotypic drug tolerance. The guinea pig remains an excellent model to test the efficacy of new drugs and drug combinations and to test the safety and efficacy of experimental vaccines intended for use in humans. Like humans, guinea pigs are among the few animals that lack the enzyme L-gulonolactone oxidase (GLUO), which is essential for the synthesis of the critical antioxidant and vitamin cofactor ascorbic acid (vitamin C). Guinea pigs, like humans,

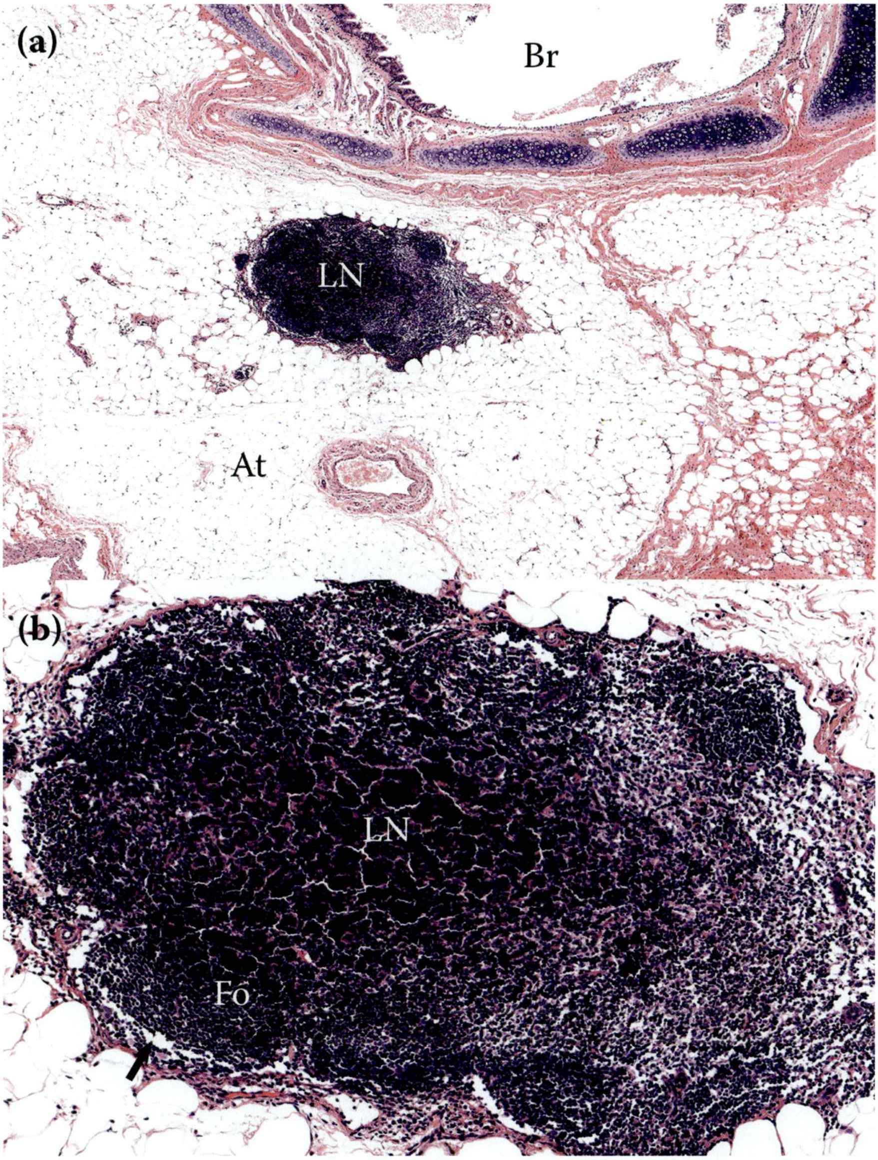

FIGURE 8.13 Normal peribronchial lymph nodes are embedded within the supportive connective and adipose tissue (H&E, LP, and MP). (a) Lymph nodes (LN) adjacent to large bronchi (Br) are embedded within a connective tissue stroma that is rich in adipose tissue (At). (b) A higher magnification of peribronchial lymph node (LN) illustrates normal follicles (Fo) and subcapsular sinuses (arrow).

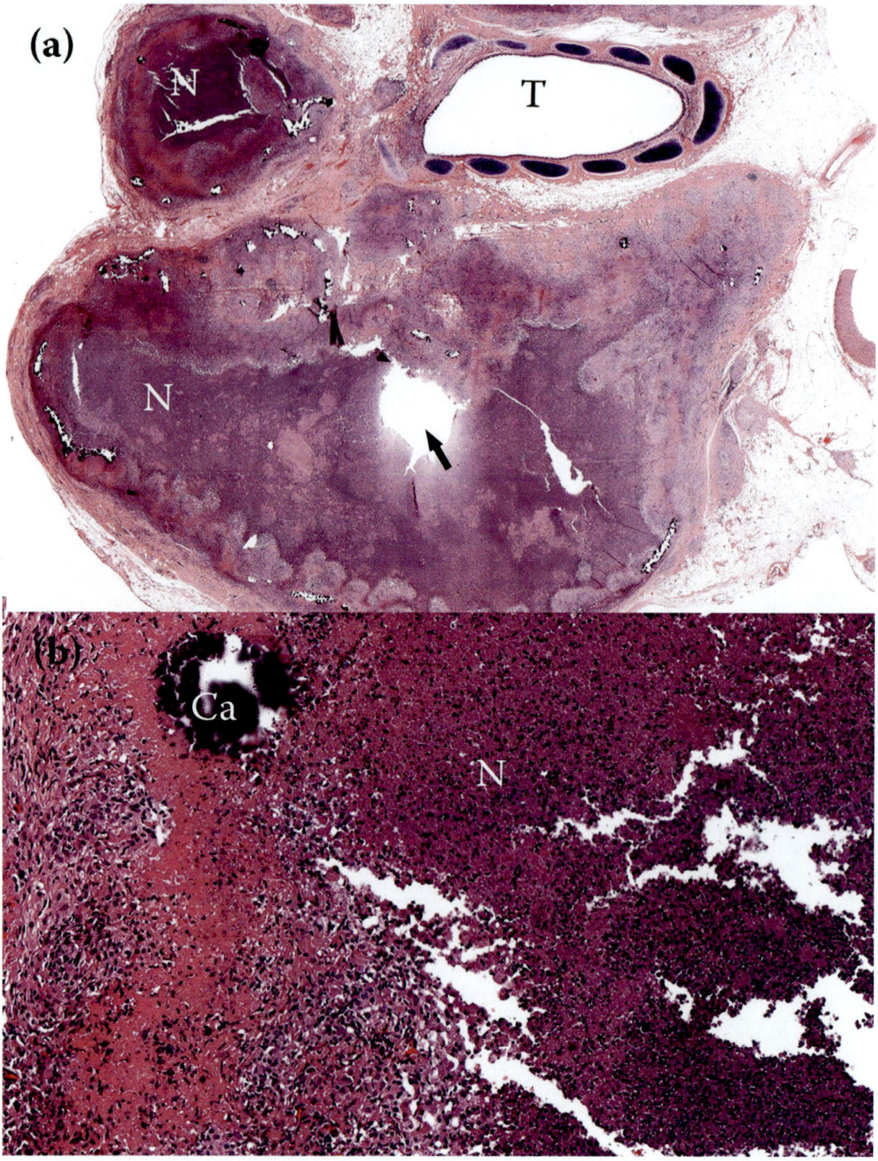

FIGURE 8.14 Extensive lymph node inflammation and necrosis is a component of early extrapulmonary dissemination in MTB infection of guinea pigs (H&E, LP, and MP). (a) LP. Granulomatous inflammation of draining lymph nodes adjacent to large airways and the trachea (T) progresses rapidly to extensive necrosis (N) and occasionally liquefactive necrosis leading to cavitation (arrow). (b) MP. Lymph node necrosis (N) with foci of calcification (Ca).

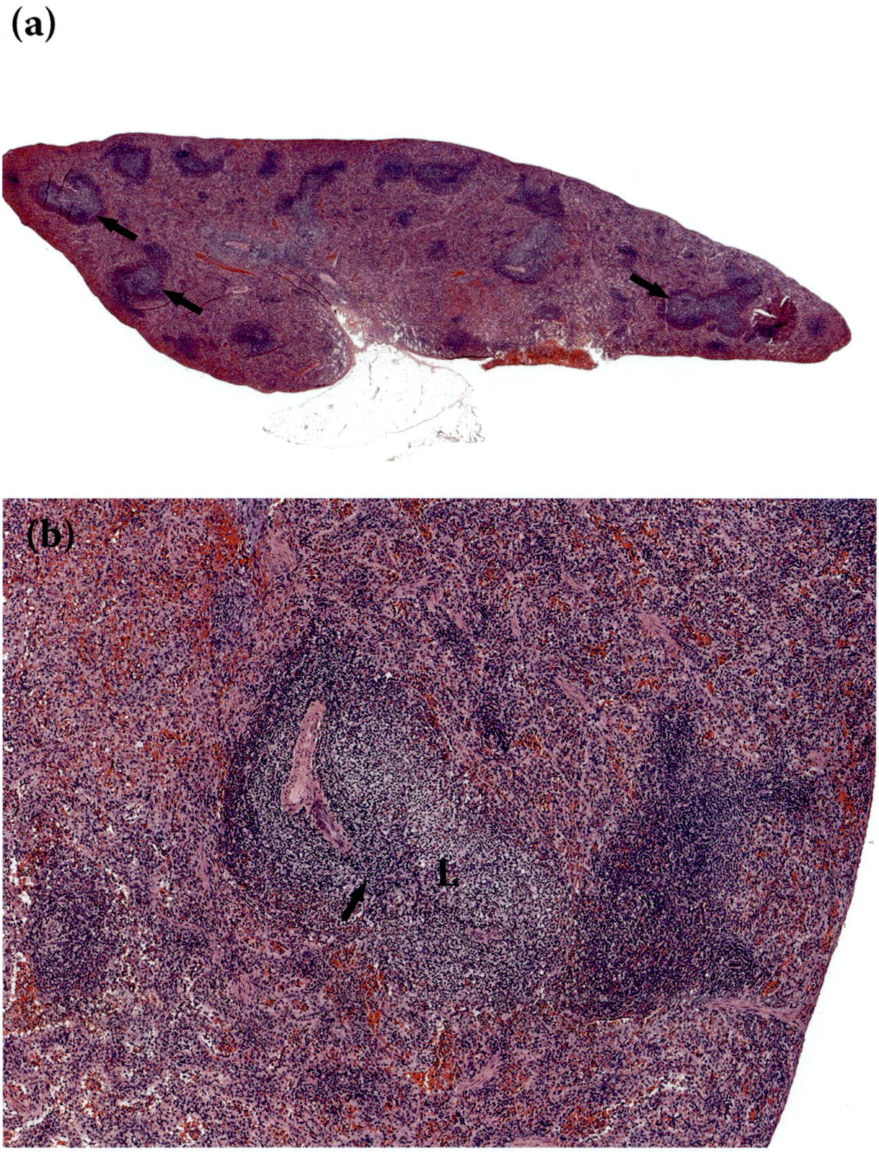

FIGURE 8.15 The normal anatomic features of the guinea pig spleen include lymphoid follicles associated with periarteriolar lymphatic sheaths (H&E, LP, and MP). (a) Low magnification of spleen shows multiple prominent periarteriolar lymphatic sheaths (arrows). (b) Higher magnification of normal spleen shows a single periarteriolar sheath (arrow) composed of mature lymphocytes (L).

(a)

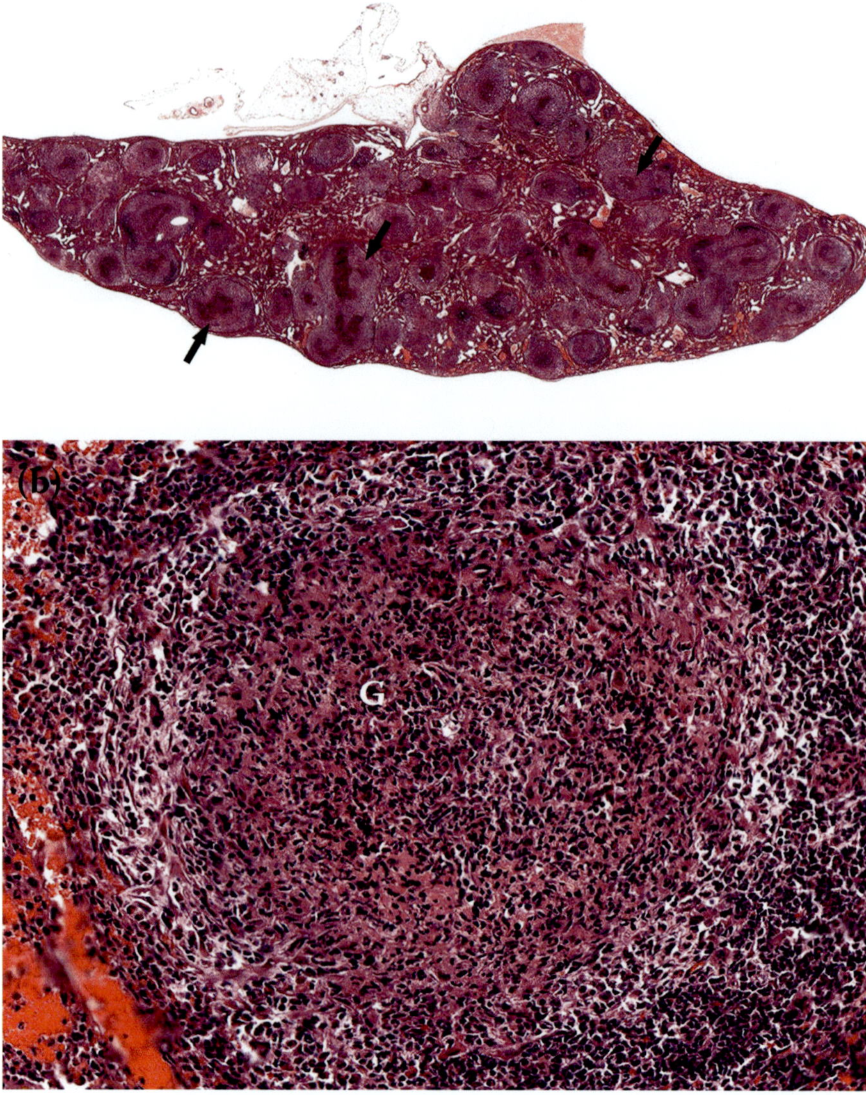

FIGURE 8.16 In the chronic stages of MTB infection in guinea pigs, spleen is among the common organs affected by extrapulmonary dissemination of bacilli (H&E, LP, and HP). (a) LP. Hematogenous dissemination of bacilli to the spleen results in multiple granulomas that often develop central necrosis (arrows). (b) HP. Higher magnification of a periarteriolar lymphatic sheath in the spleen with a centrally located granuloma (G) showing early signs of individual cell necrosis.

require ascorbic acid as a daily dietary supplement. The model therefore is of value to investigate general malnutrition and specifically vitamin C deficiency as risk factors for tuberculosis (McMurray et al., 1985; Cohen et al., 1987; McMurray et al., 1990).

SHORTCOMINGS

The primary limitation of the guinea pig model is that it lacks the diversity of lesion types that is seen in the naturally occurring disease in humans, particularly lesions with cavitation. While cavitary lesions occasionally develop spontaneously following experimental MTB infection, a method to reliably reproduce this important lesion type has not been described.

REFERENCES

Balasubramanian, V., Guo-Zhi, W., Wiegeshaus, E., and Smith, D. 1992a. Virulence of *Mycobacterium tuberculosis* for guinea pigs: A quantitative modification of the assay developed by Mitchison. *Tuber Lung Dis,* 73, 268–272.

Balasubramanian, V., Wiegeshaus, E.H., and Smith, D.W. 1992b. Growth characteristics of recent sputum isolates of *Mycobacterium tuberculosis* in guinea pigs infected by the respiratory route. *Infect Immun,* 60, 4762–4767.

Balasubramanian, V., Wiegeshaus, E.H., and Smith, D.W. 1994. Mycobacterial infection in guinea pigs. *Immunobiology,* 191, 395–401.

Basaraba, R.J. 2008. Experimental tuberculosis: The role of comparative pathology in the discovery of improved tuberculosis treatment strategies. *Tuberculosis (Edinb),* 88 Suppl. 1, S35–47.

Basaraba, R.J., Dailey, D.D., McFarland, C.T., Shanley, C.A., Smith, E.E., McMurray, D.N., and Orme, I.M. 2006a. Lymphadenitis as a major element of disease in the guinea pig model of tuberculosis. *Tuberculosis (Edinb),* 86, 386–394.

Basaraba, R.J., Smith, E.E., Shanley, C.A., and Orme, I.M. 2006b. Pulmonary lymphatics are primary sites of *Mycobacterium tuberculosis* infection in guinea pigs infected by aerosol. *Infect Immun,* 74, 5397–5401.

Cohen, M.K., Bartow, R.A., Mintzer, C.L., and McMurray, D.N. 1987. Effects of diet and genetics on *Mycobacterium bovis* BCG vaccine efficacy in inbred guinea pigs. *Infect Immun,* 55, 314–319.

Fox, W. 1868. *On the artificial production of tubercle in the lower animals.* London: Macmillan and Co.

Horwitz, M.A., and Harth, G. 2003. A new vaccine against tuberculosis affords greater survival after challenge than the current vaccine in the guinea pig model of pulmonary tuberculosis. *Infect Immun,* 71, 1672–1679.

Klein, E.E. 1873. *The anatomy of the lymphatic system: Volume I. The serous membranes.* London: Smith, Elder, & Co.

Klein, E.E. 1874. Contributions to the normal and pathological anatomy of the lymphatic system of the lungs. *Proceedings of the Royal Society of London, 22* (1873–1874), 133–145.

Klein, E.E. 1875. *The anatomy of the lymphatic system: Volume II. The lung.* London: Smith, Elder, & Co.

Koch, R., Brock, T.D., and Fred, E.B. 1982. The etiology of tuberculosis. *Rev Infect Dis,* 4, 4.

Lenaerts, A.J., Hoff, D., Aly, S., Ehlers, S., Andries, K., Cantarero, L., Orme, I.M., and Basaraba, R.J. 2007. Location of persisting mycobacteria in a guinea pig model of tuberculosis revealed by R207910. *Antimicrob Agents Chemother,* 51, 3338–3345.

McMurray, D.N. 2001. Disease model: Pulmonary tuberculosis. *Trends Mol Med,* 7, 135–137.

McMurray, D.N. 2003. Hematogenous reseeding of the lung in low-dose, aerosol-infected guinea pigs: Unique features of the host–pathogen interface in secondary tubercles. *Tuberculosis (Edinb),* 83, 131–134.

McMurray, D.N., Bartow, R.A., and Mintzer, C. 1990. Malnutrition-induced impairment of resistance against experimental pulmonary tuberculosis. *Prog Clin Biol Res,* 325, 403–412.

McMurray, D.N., Carlomagno, M.A., Mintzer, C.L., and Tetzlaff, C.L. 1985. *Mycobacterium bovis* BCG vaccine fails to protect protein-deficient guinea pigs against respiratory challenge with virulent *Mycobacterium tuberculosis. Infect Immun,* 50, 555–559.

Ordway, D., Henao-Tamayo, M., Shanley, C., Smith, E.E., Palanisamy, G., Wang, B., Basaraba, R.J., and Orme, I.M. 2008. Influence of *Mycobacterium bovis* BCG vaccination on cellular immune response of guinea pigs challenged with *Mycobacterium tuberculosis. Clin Vaccine Immunol,* 15, 1248–1258.

Ordway, D., Palanisamy, G., Henao-Tamayo, M., Smith, E.E., Shanley, C., Orme, I.M., and Basaraba, R.J. 2007. The cellular immune response to *Mycobacterium tuberculosis* infection in the guinea pig. *J Immunol,* 179, 2532–2541.

Orme, I.M. 2005. Tuberculosis vaccines: Current progress. *Drugs,* 65, 2437–2444.

Orme, I.M. 2006. Preclinical testing of new vaccines for tuberculosis: A comprehensive review. *Vaccine,* 24, 2–19.

Palanisamy, G.S., Smith, E.E., Shanley, C.A., Ordway, D.J., Orme, I.M., Basaraba, R.J. 2008. Disseminated disease severity as a measure of virulence of *Mycobacterium tuberculosis* in the guinea pig model. *Tuberculosis (Edinb),* 88, 295–306.

Sanderson, J.B. 1867. Sanderson; On the communicability of tubercle by inoculation. *Report of the Medical Officer.*

Schreider, J.P. 1983. *Nasal airway anatomy and inhalation deposition in experimental animals and people.* Boca Raton, FL: CRC Press.

Smith, D.W., Balasubramanian, V., and Wiegeshaus, E. 1991. A guinea pig model of experimental airborne tuberculosis for evaluation of the response to chemotherapy: The effect on bacilli in the initial phase of treatment. *Tubercle,* 72, 223–231.

Smith, D.W., and Harding, G.E. 1977. Animal model of human disease. Pulmonary tuberculosis. Animal model: Experimental airborne tuberculosis in the guinea pig. *Am J Pathol,* 89, 273–276.

Smith, D.W., and Wiegeshaus, E.H. 1989. What animal models can teach us about the pathogenesis of tuberculosis in humans. *Rev Infect Dis,* 11 Suppl. 2, S385–393.

Tyler, W.S., and Julian, M.D. 1992. *Gross and subgross anatomy of lungs, pleura, connective tissue septa, distal airways, and structural units.* Boca Raton, FL: CRC Press.

Williams, A., Hall, Y., and Orme, I.M. 2009. Evaluation of new vaccines for tuberculosis in the guinea pig model. *Tuberculosis (Edinb),* 89, 389–397.

CHAPTER 9

Pulmonary Tuberculosis in the Rat

Amit Singhal, El Moukhtar Aliouat, Colette Creusy, Gilla Kaplan, and Pablo Bifani

CONTENTS

INTRODUCTION

In the past the rat has been reported to be highly resistant to *Mycobacterium tuberculosis* (MTB) and thus unattractive as a model of experimental tuberculosis. Early studies showed that high doses of MTB could neither produce tuberculous lesions in the lung nor kill the rats (Gloyne and Page, 1923; Ornstein and Steinbach, 1925). These observations were followed by a number of studies describing the presence of nonnecrotic tubercles in the lungs of infected rats in association with cutaneous delayed type hypersensitivity (DTH) responses to tuberculin (Long and Vorwald, 1930; Hehre and Freund, 1939). In 1941 Wessels reported that the tuberculosis bacilli could disseminate to different organs in the rat and induced DTH in infected animals (Wessels, 1941a,b). In the 1960s rats were reported to be as susceptible to MTB as some strains of mice (Gray, 1961; Flax and Waksman, 1962), and in the 1970s Lefford and colleagues showed that MTB-infected rats could generate a specific immune response (Lefford et al., 1973). More recently, studies on rats addressed the kinetics of the bacillary load and the histologic response in the lungs (Sugawara et al., 2004a,b, 2006). The following chapter summarizes some of our new microbiological and histological observations in the Wistar rat tuberculosis model.

INOCULUM AND ROUTE OF INOCULATION

Rats

Specific pathogen free (SPF) outbred female Wistar rats (Crl:WI, 6 to 8 weeks old) were used. The animals were housed in isolator cages (Ingenia, France) under Biosafety Level 3 (BSL-3) conditions. Throughout the experiments animals were fed sterile irradiated food and sterile water *ad libitum*. All animals were maintained in accordance with protocols approved by the institutional animal ethical committee, and experiments were carried out in accordance with BSL-3 protocols and animal legislation.

Bacterial Strains and Inocula Preparation

Two well-characterized clinical isolates of MTB Beijing strains, W4 and HN878 (Bifani et al., 1999; Manca et al., 2001), were used in the experiments. The strains were cultured in stationary Sauton medium in 500-ml cellular culture flasks to mid-log phase, washed, centrifuged, filtered to remove clumps, and adjusted in PBS/glycerol to an OD_{600} of 1.0 corresponding to approximately 2×10^8 colony forming units/milliliter (CFUs/ml). The bacillary suspension was dispensed into small aliquots and stored at –80°C, thawed one day later, and plated on 7H11 medium for CFU enumeration. On the day of infection the bacillary suspension was serially diluted in PBS, and rats were infected with inocula varying from 10^2 to 10^5 bacteria.

Infection Procedure

Rats were infected by an in-house-developed nonsurgical endotracheal procedure (Garry et al., 2003). Briefly, animals were anesthetized intraperitoneally with a cocktail consisting of ketamine: 75 mg/kg; diazepam: 1.2 mg/kg; and atropine: 0.4 mg/kg. A cold light, a blunt rod, an opening mouth device and a catheter device (SURFLO® i.v. catheter 16G × 2," Terumo®, France) were utilized. For the bacterial inoculation, the anesthetized rat was suspended by its upper incisors on a wire rod at the top of a Teflon slant board (16 cm by 25 cm; 60 degree angle). The mouth of the rat was opened, and the cold light was placed in front of the neck area. The tongue was gently stretched outwards, exposing a bright spot at the back of the throat corresponding to the illumination of the opening of the trachea; which was easily seen through the open glottis. A guiding blunt rod was introduced inside the trachea and used to thread a catheter. The rod was removed once the catheter was in place. In order to ensure that the catheter was inside the trachea and not the esophagus, soap foam was introduced at the opening of a 5-ml syringe (without the piston) and connected with the catheter. The regular movement of the soap on the upper part of the syringe indicates proper respiration and hence correct placement of the catheter. A suspension of 0.1 ml of the MTB inoculum was administered into the trachea followed by 0.9 ml of air to disseminate the bacilli within the lung by connecting a needless syringe to the catheter. To optimize the dispersion of bacterial inoculum, a second injection of 1 ml of air was administered.

Histopathological Evaluation of the Lungs

At predetermined time points postinfection (14 to 180 days), MTB-infected rats were euthanized with sodium pentobarbital (55 mg/kg) administered intraperitoneally. The lungs were excised, washed with PBS, and fixed with 10% neutral formalin (for 4 days), embedded in paraffin, and processed for histopathological examination. Sections (5 μm) were stained with (1) hematoxylin-eosin-safran (H&E), (2) Sirius red and/or Volgens-Gomori (collagen types I and III, for analysis of fibrosis), (3) Ziehl-Neelsen (ZN) for acid-fast bacilli (AFB). Light microscopy was performed on Leica DMRB microscope.

FEATURES

Wistar Rats Control Growth of Tubercle Bacilli

In low dose-infected Wistar rats (<10^3 bacilli), the lung bacterial load peaked at day 21 to 28 postinfection followed by a gradual decrease in bacillary numbers over the next 120 to 180 days (Figure 9.1). The course of infection in animals inoculated with a high bacillary dose (>10^3) peaked similarly by 21 to 28 days postinfection, but then stabilized in the lungs over the next 120 to 180 days (Figure 9.1). The extent of bacillary control was also affected to some extent by the strain of MTB used for the infection.

Lung Histopathology

Overall, the pathology observed in the lungs of MTB infected Wistar rats was dose dependent. Lung histopathology over the course of infection revealed an early increase in the size and number of granulomatous cellular accumulation followed by a gradual resorption of the lesions, which correlates with a decrease in lung CFUs. Thickening of the alveolar walls and influx of inflammatory cells without any granulomatous lesions was observed from days 14 to 60 in the lungs of rats infected with a low dose (<10^3 bacilli). In contrast, granulomas containing lymphocytes, epithelioid, and Langhans' cells were clearly visible in the lungs and lymph nodes of rats infected with >10^3 bacteria from day 14 onward. In the rest of

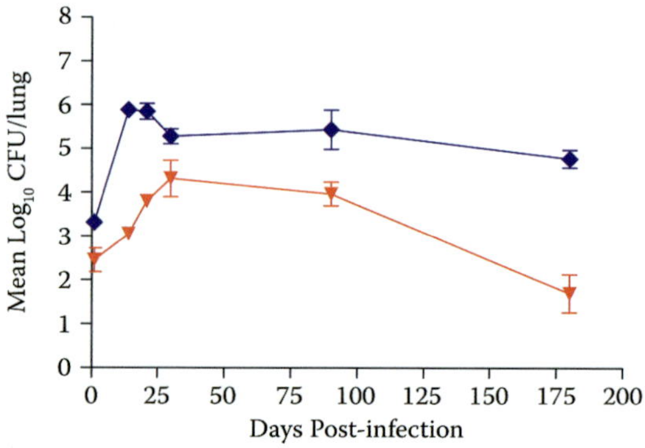

FIGURE 9.1 Bacterial load in the lungs of Wistar rats infected with MTB W4 strain. Growth of mycobacteria in the lungs of Wistar rats upon infection with different inocula of bacilli. All results are expressed logarithmically as the mean $\log_{10}$ bacilli CFU ± SD (n = 4 to 12 rats per point). Rats were infected by 2.4 ± 0.4 and 3.3 ± 0.1 $\log_{10}$CFU (red triangles and blue diamonds respectively) as determined after 24 hours of infection.

this section, we define the acute phase as the first 28 days of infection, the chronic phase expanding to the 90th day postinfection, and the quiescent stage from day 90 to 180 postinfection. Figure 9.2 shows the structure of healthy Wistar rat lung.

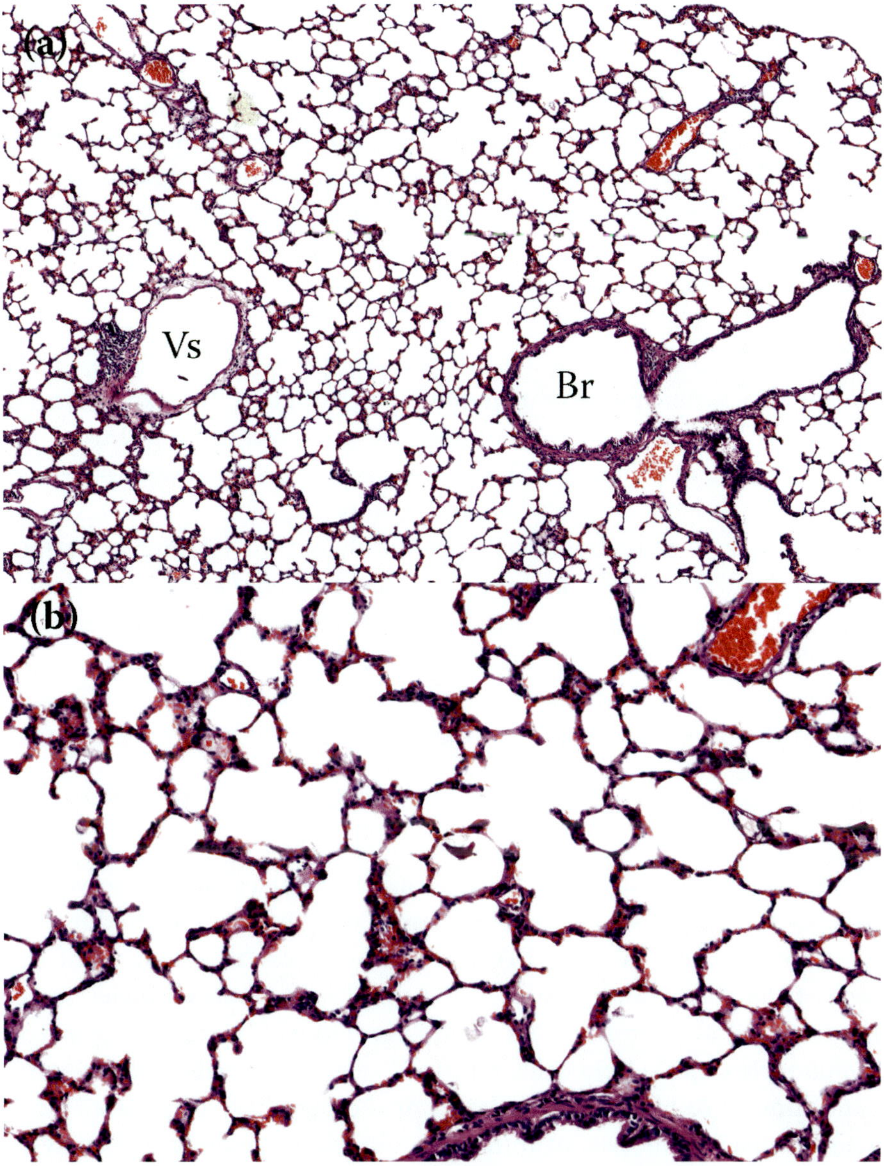

FIGURE 9.2 Normal Wistar rat lung (H&E, LP, and HP). (a) Low power demonstrating bronchiole (Br) and blood vessel (Vs). (b) HP.

Acute (Early)

Histopathological examination of the lungs of rats infected with a high dose (>10^3 bacilli) evaluated at 14 days postinfection revealed well-defined inflammatory foci (Figure 9.3a), an increase of peribronchiolovascular inflammatory infiltrate (Figure 9.3b) and early stage granulomas comprising lymphocytes, clusters of macrophages, and some multinucleated generic giant cells (Figure 9.3c). Ziehl-Neelsen staining revealed many intracellular acid-fast bacilli (Figure 9.3d). By the 28th day postinfection different types of lesions at various stages of development were observed. These included early stage granulomas and more defined solid granulomatous lesions with epithelioid cells mixed with numerous lymphocytes (Figure 9.4). The granulomas ranged from 0.5 to 1 mm^2 in size and lacked any necrosis and fibrosis.

Chronic (Late)

During the chronic phase of infection by the 60th day, the granulomas in the lungs of rats infected with a high dose (>10^3 bacilli) ranged from 1.0 to 1.5 mm^2 in size (Figure 9.5). Different types of lesions were observed including nonnecrotizing granulomas comprised of a central cluster of densely packed epithelioid histiocytes with occasional multinucleated giant cells surrounded by a crown of lymphocytes but no necrotic debris (Figure 9.5a), and subpleural granulomas having lymphocytes mixed with epithelioid histiocytes surrounded by foamy macrophages (Figure 9.5b). By the 90th day postinfection the presence of foamy macrophages and Langhans' type giant cells at the inflammatory foci became more apparent (Figure 9.6).

Quiescent

In low dose-infected Wistar rats, a quiescent phase was observed in some of the animals whereby few if any CFU could be cultured by 180 days postinfection (Figure 9.1). Histopathological examination of the lungs from these rats revealed some small granulomas/follicles ranging from 0.2 to 0.4 mm^2 in size. These granulomas consisted of mixed lymphocytes, epithelioid cells, multinucleated giant cells, and a significant number of foamy macrophages at the periphery (Figure 9.7). No acid-fast bacilli, fibrosis, or mineralization could be detected in any of the animals at this stage. Multinucleated giant cells were also observed in lymph nodes of some of these animals (Figure 9.7d). In contrast to

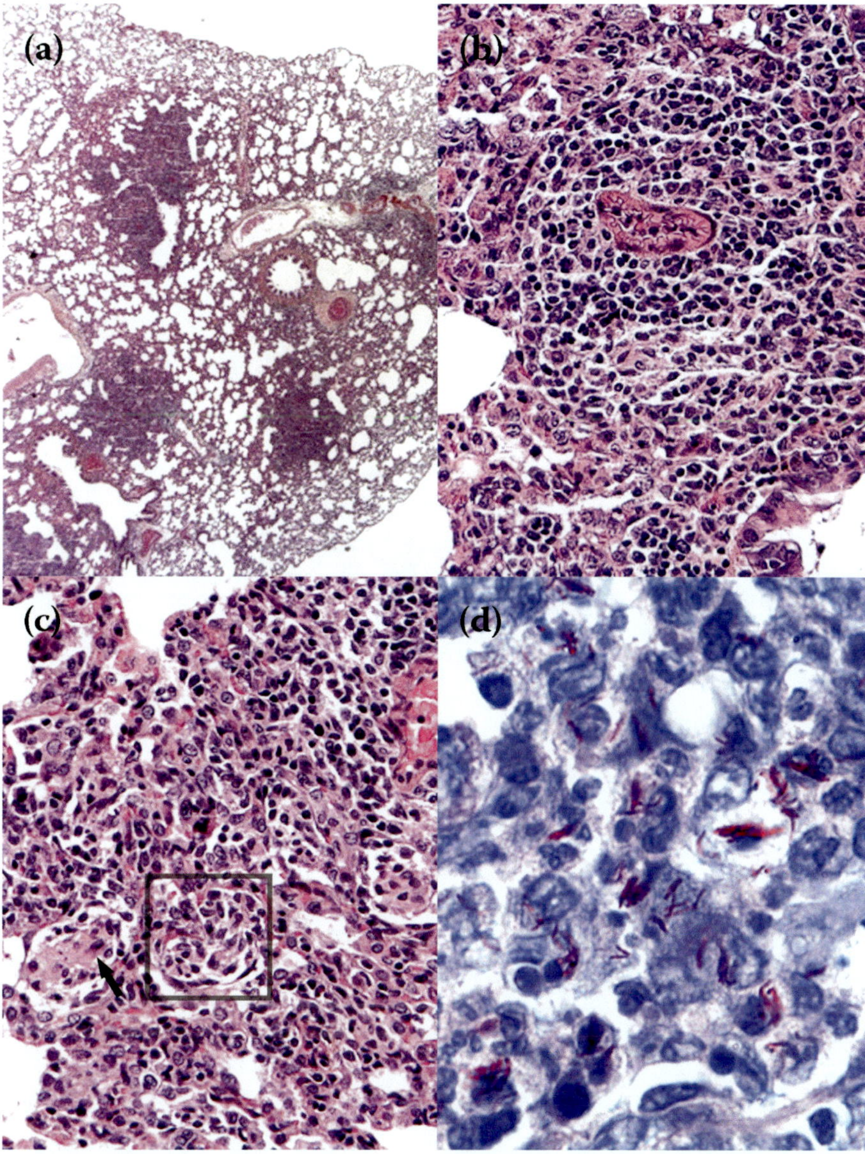

FIGURE 9.3 Wistar rat lung, day 14 postinfection with high-dose (>10^3 bacilli) of MTB. Representative photomicrographs from sections of formalin-fixed and paraffin-embedded lung tissue collected on day 14 postinfection are shown. (a) H&E, LP. Well-defined inflammatory foci. (b) H&E, HP. Important increase of peribronchiolovascular inflammatory infiltrate encompassing mainly lymphocytes and plasma cells. (c) H&E, HP. Early stage granulomas with cluster of macrophages (square area) intermixed with lymphocytes and giant cell (arrow). (d) ZN oil, HP. Numerous acid-fast bacilli, both intra- and extracellular.

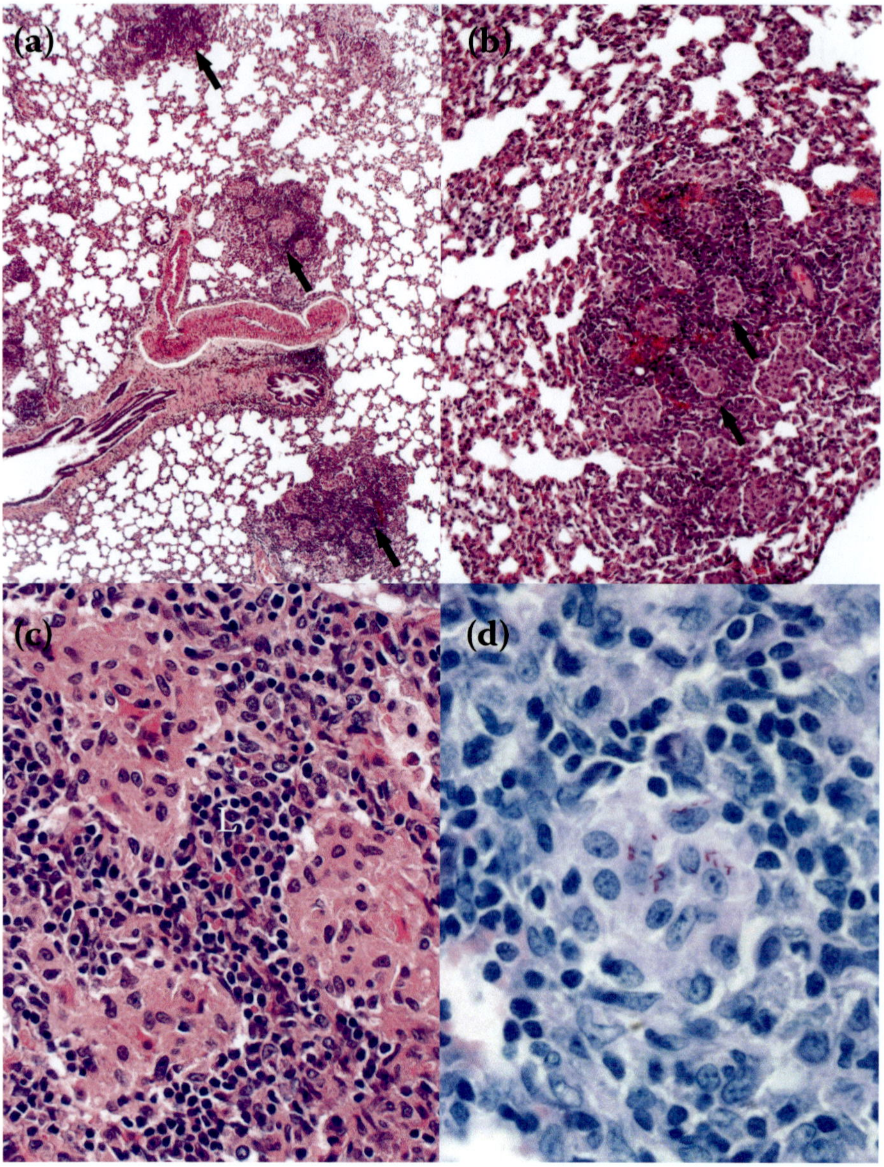

FIGURE 9.4 Wistar rat lung, day 28 postinfection with high-dose (>10^3 bacilli) of MTB. Representative photomicrographs from sections of formalin-fixed and paraffin-embedded lung tissue collected on day 28 postinfection are shown. (a) and (b) Granulomas at different stages of development (arrows) with varying degrees of consolidation (H&E, LP). (c) H&E, HP. Nonnecrotizing granulomas with mixed epithelioid cells and lymphocytes (L). (d) ZN oil, HP. Small granuloma demonstrating multiple acid-fast bacilli (magenta).

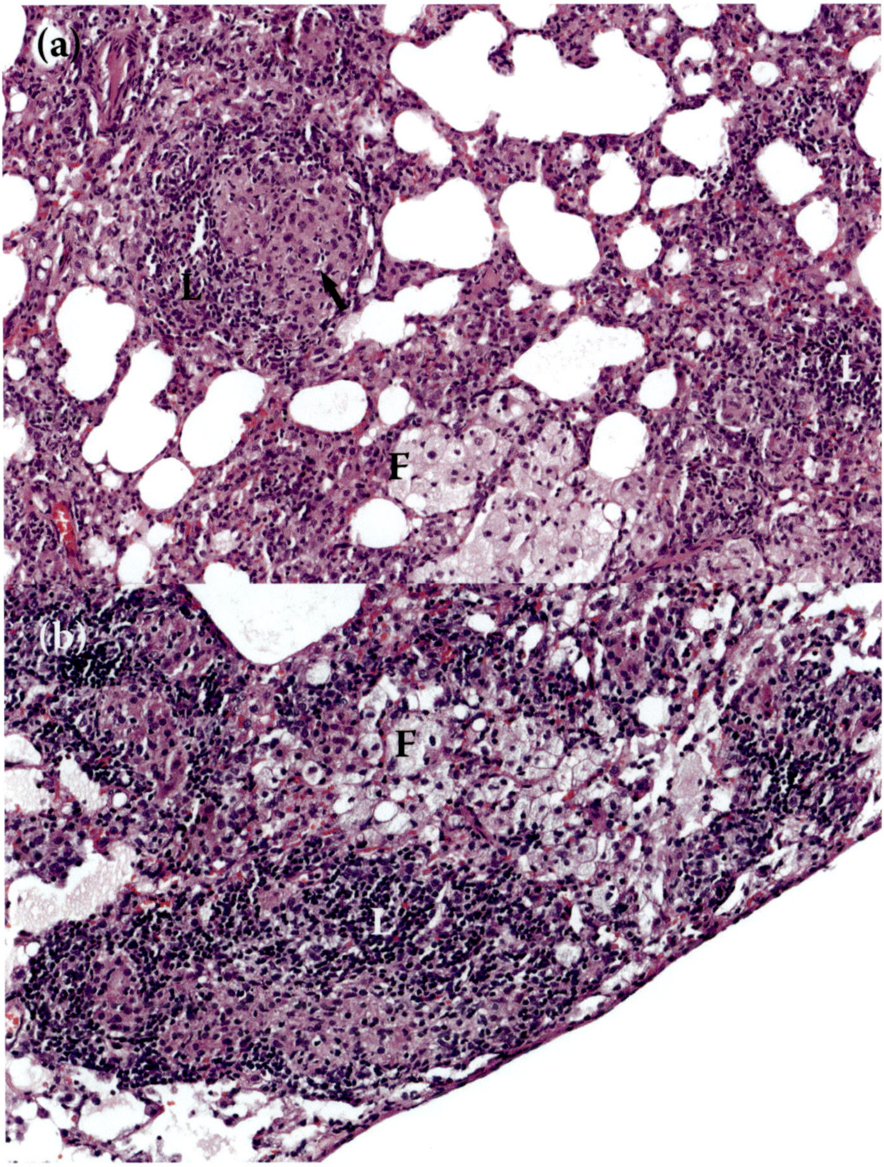

FIGURE 9.5 Wistar rat lung, day 60 postinfection with high-dose (>10^3 bacilli) of MTB (H&E). (a) MP. Nonnecrotizing granuloma with a central cluster of epithelioid cells (arrow) surrounded by lymphocytes (L) and loose foamy macrophage aggregate (F). (b) MP. Subpleural mixed cell infiltrate with mixed epithelioid cells, lymphocytes (L), and foamy macrophages (F).

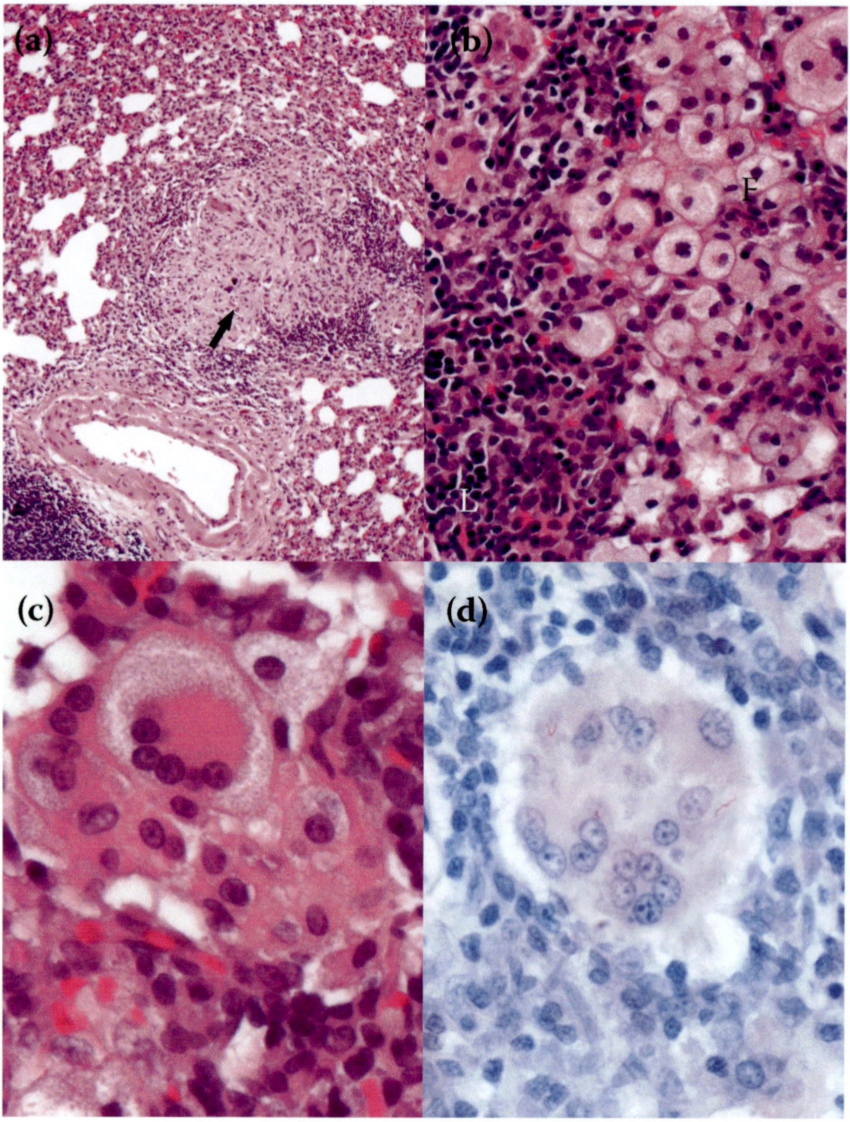

FIGURE 9.6 Foamy macrophages and multinucleate giant cells, day 90 postinfection with high-dose (>10^3 bacilli) of MTB. (a) H&E, LP. Nonnecrotizing granuloma (arrow) with focal Langhans' type multinucleate giant cell and peripheral lymphoplasmacytic aggregates. (b) H&E, MP. Collections of foamy macrophages (F) are scattered throughout the inflamed lung and mix with lymphocytes and other chronic inflammatory cells (L). Foam cells are common in areas of inflammation or cellular damage, and their appearances are the result of phagocytosis of lipids from injured cells. (c) H&E, HP. Langhans' type multinucleate giant cell. (d) ZN oil, HP. Foreign-body type multinucleate giant cell with several acid-fast bacilli (magenta).

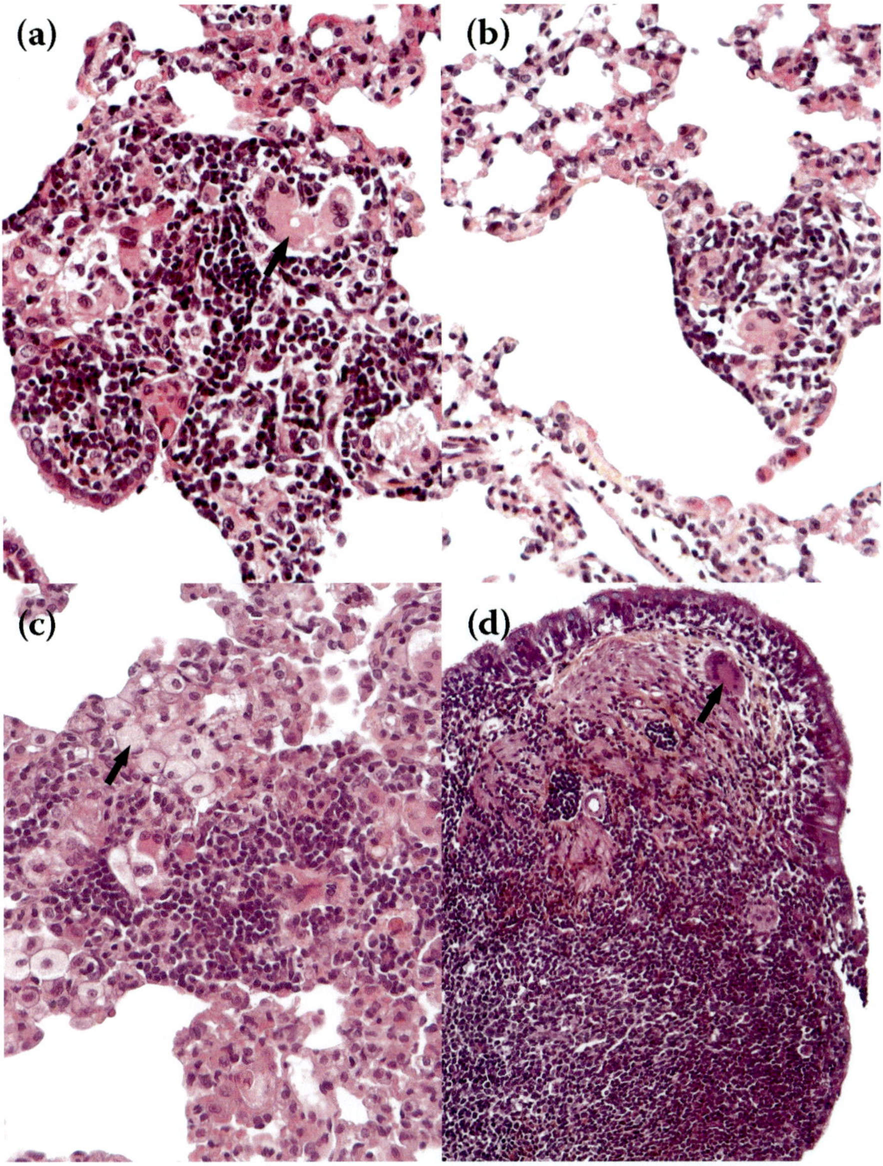

FIGURE 9.7 Wistar rat lung, day 180 postinfection with low-dose (<10^3 bacilli) of MTB (H&E, HP, and LP). (a) and (b) HP. Small cellular aggregate with mixed epithelioid cells, lymphocytes, and multinucleated Langhans' cell (arrow). (c) HP. Numerous peripheral foamy macrophages (arrow) were found in close proximity to the granuloma. (d) LP. Nonnecrotizing granuloma with focal Langhans' type multinucleated giant cell (arrow) and peripheral lymphocyte aggregates in bronchus associated lymphoid tissue.

the low dose–infected rats, the lungs of high dose-infected animals had numerous bacteria at 180 days postinfection (Figure 9.1). The lungs had many peribronchiolovascular granulomas ranging from 0.3 to 0.8 mm^2 in size (Figure 9.8a,b). Some of these lesions contained more macrophages relative to lymphocytes and cholesterol clefts at the periphery (Figure 9.8c). Minimal fibrosis could be observed surrounding these granulomas (Figure 9.8d).

VARIATIONS

Necrotic Granuloma

Necrotic lesions, a hallmark of human tuberculosis, are rare in the rat model and are only observed in animals infected with high doses of tubercle bacilli (Figure 9.9). These lesions consist of a necrotic core surrounded by epithelioid histiocytes and lymphocytes in the periphery (Figure 9.9a). Distinct cellular layers forming almost concentric circles were noted (Figure 9.9b). An outer crown of lymphocytes intermixed with fibroblasts could be observed by reticulin staining (Figure 9.9c,d).

Lesions during Quiescent Stage

Animal-to-animal variations in lung histopathology were observed within the low dose-infected Wistar rats at 180 days postinfection. The appearance of lung parenchyma varied from normal (Figure 9.10a) to highly cellular (Figure 9.10b). The dense cellular areas were filled with very small follicles comprising lymphocytes, epithelioid histiocytes, and multinucleated giant cells (Figure 9.10c,d).

STRENGTHS OF THIS ANIMAL MODEL

Appearance of Quiescent Phase

The most significant observation in the Wistar rat model is the natural control of *M. tuberculosis* infection in low dose–infected rats, a phenotype which is observed in the macaque model (Capuano et al., 2003; Lin et al., 2006), and in rabbits infected with low-dose MTB CDC1551 (Kaplan and Tsenova, see Chapter 7). The potential of the Wistar rat TB model in understanding host–pathogen interactions during the subclinical stage of disease needs to be further investigated.

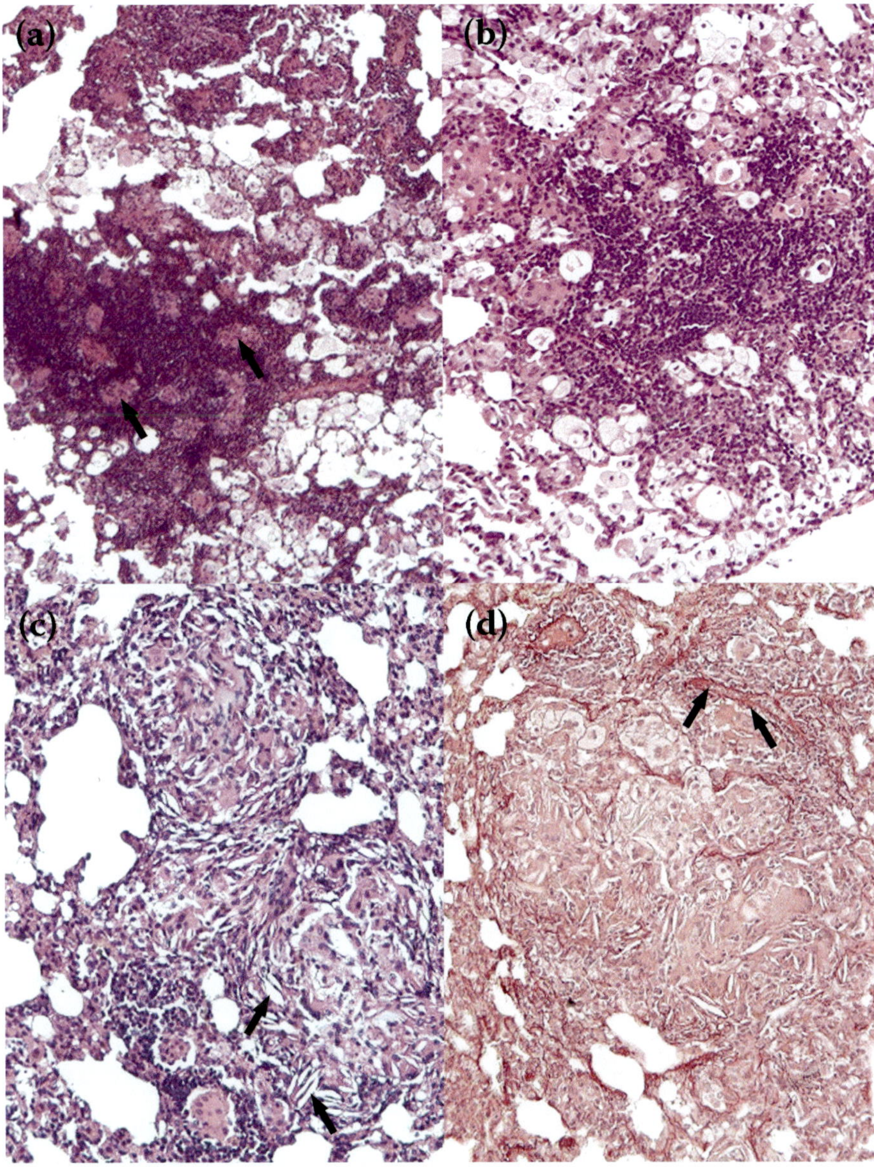

FIGURE 9.8 Wistar rat lung, day 180 postinfection with high-dose (>10^3 bacilli) of MTB. (a) and (b) H&E, MP. Numerous small peribronchiolovascular granulomas predominantly composed of epithelioid histiocytes with occasional giant cells. (c) H&E, MP. Several cholesterol clefts around some of the granulomas (arrows). (d) Sirius red staining, MP. Mild fibrosis (arrows) around the granulomas.

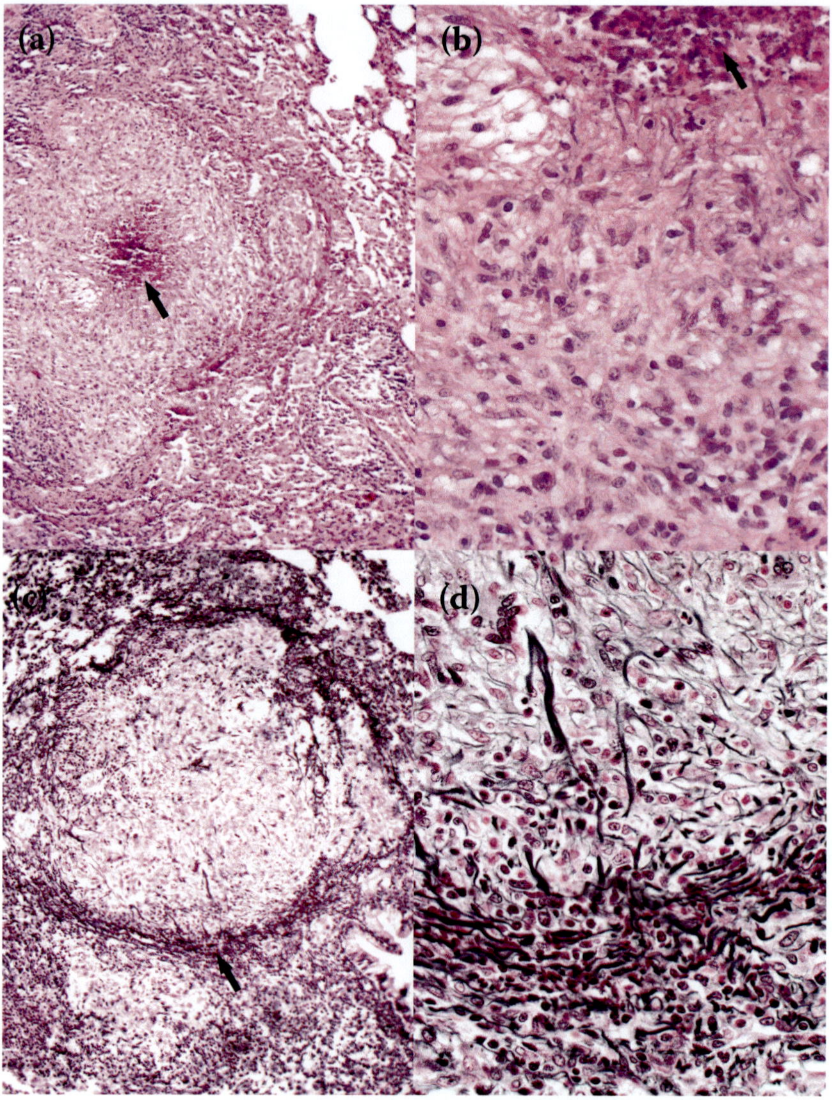

FIGURE 9.9 Necrotic granuloma at day 60 postinfection with high dose (>10^3 bacilli) of MTB. (a) H&E, LP. Low-power view demonstrating a necrotizing granuloma composed of plump aggregates of epithelioid histiocytes surrounding a necrotic core (arrow). A suggestion of peripheral fibrosis is evident at this magnification. (b) H&E, HP. At high power the area of necrosis (arrow) is in sharp contrast to the epithelioid histiocytes, which have vesicular nuclei, abundant pale cytoplasm, and indistinct cellular borders. (c) Reticulin staining (black) demonstrates the development of a fibrocollagenous outer later (LP). (d) High-power view of the edge of the granuloma shown in (c) demonstrates prominent reticulin fibers at the periphery, which also extend into the granuloma between histiocytic aggregates.

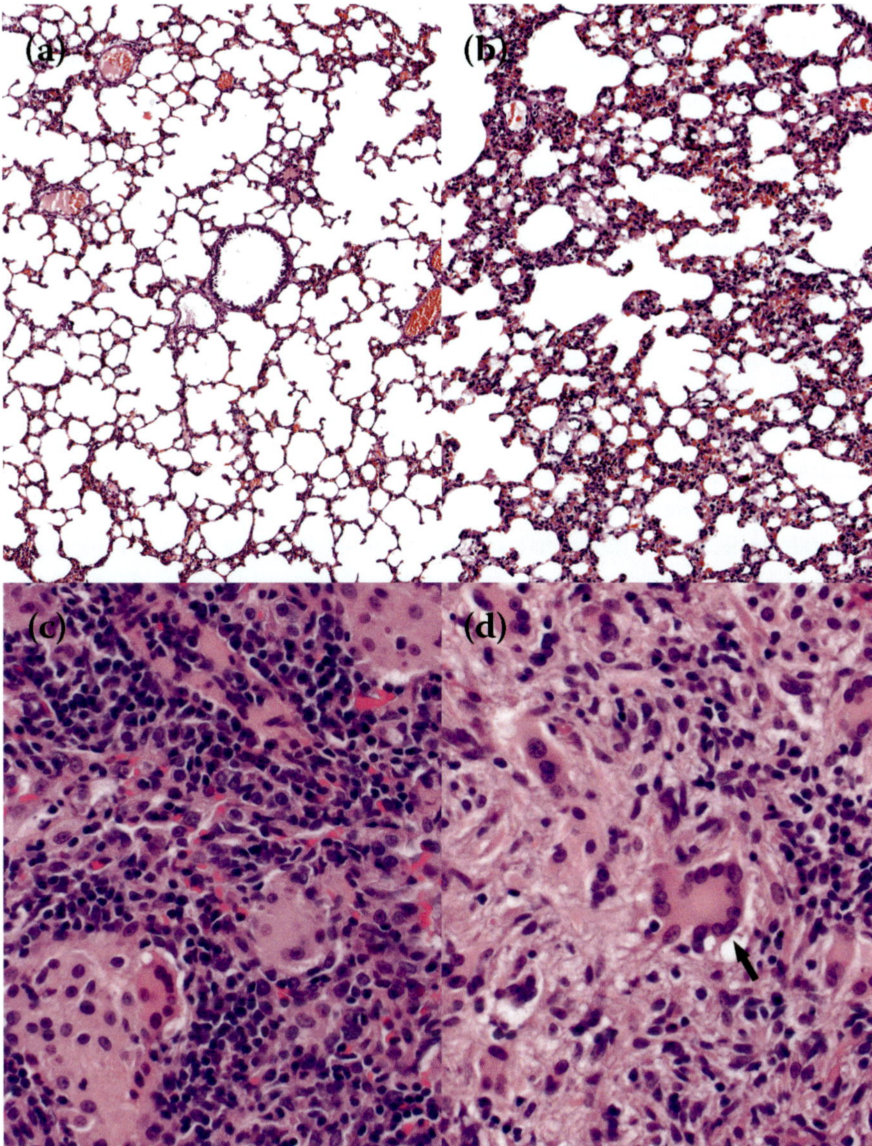

FIGURE 9.10 Subclinical phase of infection, day 180 postinfection by low-dose (<10^3 bacilli) of MTB (H&E, LP, and HP). (a) LP. Apparently normal lung parenchyma. (b) LP. Cellular lung parenchyma predominantly mononuclear interstitial infiltrate. (c) and (d) HP. Small granulomas surrounded by a lymphoplasmacytic infiltrate with multinucleated giant cells, some of Langhans' type (arrow).

Organized Lesions

The Wistar rat MTB model illustrated here is found to develop tuberculous granulomas, defined as granulomatous inflammation composed of epithelioid histiocytes surrounded by giant cells of Langhans' type, lymphocytes, and fibroblasts (Blood et al., 2007). The rat presents an additional and alternative model which can provide a wealth of knowledge in understanding the disease physiology.

Implications in Drug Discovery

A major bottleneck in the evaluation and early development of TB drug candidates is the lack of a predictive animal model that is compatible with the pragmatic needs of medicinal chemistry lead optimization programs. An effective efficacy model must be sensitive to time required for testing and to resources available for this process. Because it is relatively cheap, small, and easy to handle, the rat has been a species of choice for early drug development in the pharmaceutical industry. The rat therefore would have a clear advantage should a robust efficacy model be available, which effectively reproduces at least some properties of disease progression seen in humans. Efficacy data could be directly correlated with pharmacokinetic/pharmacodynamic characterization and with toxicology studies in the same animal model.

SHORTCOMINGS

Lack of Full Spectrum of Human Tuberculosis Histopathology

One of the major drawbacks of the Wistar rat model is that necrotic and fibrotic granulomas in the lungs of MTB-infected rats are not commonly observed. Only when rats are infected at high doses were single necrotic lesions seen ($>10^3$). Moreover, pulmonary cavity formation and mineralization, associated with human tuberculosis, have not been observed.

REFERENCES

Bifani, P.J., Mathema, B., Liu, Z., Moghazeh, S.L., Shopsin, B., Tempalski, B., Driscol, J., Frothingham, R., Musser, J.M., Alcabes, P., and Kreiswirth, B.N. 1999. Identification of a W variant outbreak of *Mycobacterium tuberculosis* via population-based molecular epidemiology. *JAMA*, 282, 2321–2327.

Blood, D.C., Studdert, V.P., and Gay, C.C. 2007. *Saunders comprehensive veterinary dictionary*, 3rd edition. W.B. Saunders, New York.

Capuano, S.V., 3rd, Croix, D.A., Pawar, S., Zinovik, A., Myers, A., Lin, P.L., Bissel, S., Fuhrman, C., Klein, E., and Flynn, J.L. 2003. Experimental *Mycobacterium tuberculosis* infection of cynomolgus macaques closely resembles the various manifestations of human *M. tuberculosis* infection. *Infect Immun,* 71, 5831–5844.

Flax, M.H., and Waksman, B.H. 1962. Delayed cutaneous reactions in the rat. *J Immunol,* 89, 496–504.

Garry, S., Nesslany, F., Aliouat, E., Haguenoer, J.M., and Marzin, D. 2003. Assessment of genotoxic effect of benzo[a]pyrene in endotracheally treated rat using the comet assay. *Mutat Res,* 534, 33–43.

Gloyne, S.R., and Page, D.S. 1923. The reaction to *B. tuberculosis* in the albino rat. *J Pathol Bacteriol,* 26, 224–233.

Gray, D.F. 1961. The relative natural resistance of rats and mice to experimental pulmonary tuberculosis. *J Hyg (Lond),* 59, 471–477.

Hehre, E., and Freund, J. 1939. Sensitization, antibody formation and lesions produced by tubercle bacilli in the albino rat. *Arch Pathol,* 27, 287–306.

Lefford, M.J., McGregor, D.D., and Mackaness, G.B. 1973. Immune response to *Mycobacterium tuberculosis* in rats. *Infect Immun,* 8, 182–189.

Lin, P.L., Pawar, S., Myers, A., Pegu, A., Fuhrman, C., Reinhart, T.A., Capuano, S.V., Klein, E., and Flynn, J.L. 2006. Early events in *Mycobacterium tuberculosis* infection in cynomolgus macaques. *Infect Immun,* 74, 3790–3803.

Long, E.R., and Vorwald, A.J. 1930. An attempt to influence the growth of the tubercle bacillus in the animal body by modifying the concentration of a growth promoting substance (glycerol) in the tissues. *Am Rev Tuberc,* 22, 636–654.

Manca, C., Tsenova, L., Bergtold, A., Freeman, S., Tovey, M., Musser, J.M., Barry, C.E., 3rd, Freedman, V.H., and Kaplan, G. 2001. Virulence of a *Mycobacterium tuberculosis* clinical isolate in mice is determined by failure to induce Th1 type immunity and is associated with induction of IFN-alpha/beta. *Proc Natl Acad Sci USA,* 98, 5752–5757.

Ornstein, G.B., and Steinbach, M.M. 1925. The resistance of the albino rat to infection with tubercle bacilli. *Am Rev Tuberc,* 12, 77–86.

Sugawara, I., Yamada, H., and Mizuno, S. 2004a. Pathological and immunological profiles of rat tuberculosis. *Int J Exp Pathol,* 85, 125–134.

Sugawara, I., Yamada, H., and Mizuno, S. 2004b. Pulmonary tuberculosis in spontaneously diabetic Goto Kakizaki rats. *Tohoku J Exp Med,* 204, 135–145.

Sugawara, I., Yamada, H., and Mizuno, S. 2006. Nude rat (F344/N-rnu) tuberculosis. *Cell Microbiol,* 8, 661–667.

Wessels, C.C. 1941a. Tuberculosis in the rat. I. Gross organ changes and tuberculin sensitivity in rats infected with tubercle bacilli. *Am Rev Tuberc,* 43, 449–458.

Wessels, C.C. 1941b. Tuberculosis in the rat. II. The fate of tubercle bacilli in the various organs of the rat. *Am Rev Tuberc,* 43, 459–474.

CHAPTER 10

Pulmonary Tuberculosis in the Mouse

Sowmya Bharath and V. Balasubramanian

CONTENTS

INTRODUCTION

Mouse, *Mus musculus,* has now become the proverbial guinea pig for most pharmacological investigations. Tuberculosis is no exception. To a very large extent, the insights into the pathogenesis of tuberculosis, especially with respect to genetic susceptibility, immune response in early and late stages, have been gained from the studies involving the large library of genetically defined inbred strains of mice. The analysis of most aspects of human tuberculosis is possible in the mouse model, due to the availability of hundreds of inbred, congenic, recombinant, mutant, and genetically engineered strains and an abundance of immunological reagents. The humble mouse scores over other laboratory animal species, simply because such an assortment is far more limited for studies in other animal species. In addition, experiments in mice are far less expensive as compared to other species (McMurray et al., 1996; Dharmadhikari and Nardell, 2008).

While any one particular strain of mice may not fully capture the immunopathology of the entire spectrum of human tuberculosis, the combined information from various genetic backgrounds would capture the spectrum adequately. Apt and Kramnik (Apt and Kramnik, 2009) in their recent commentary have pointed out that the spectrum of human tuberculosis in terms of pathophysiology can be explained due to the large degree of heterogeneity in the human population, and that the lack of a spectrum of pathophysiological phenotypes in mice is simply due to the fact that one is focused to one inbred strain or the other. Against this background, it is important to note that the present set of histopathological observations is from only one strain of mice, namely BALB/c, and from a single set of infection conditions. Therefore, it is not unexpected that the full gamut of histopathological findings associated with human tuberculosis will not be seen in this section (Figure 10.1 and Figure 10.2).

FIGURE 10.1 (opposite page) Normal histology (H&E). The conducting airways of mouse lung terminate abruptly at noncartilaginous and nonalveolarized terminal bronchioles, which open directly into alveolar ducts. These in turn communicate with alveoli. The alveoli comprise two types of epithelial cells and cells of the reticuloendothelial system. Other cell types which are part of normal histology include alveolar histiocytes and lymphocytes. (a) This medium-power view demonstrates blood vessels (Vs), alveolar duct and alveoli (Al), and bronchiole (Br) lined by

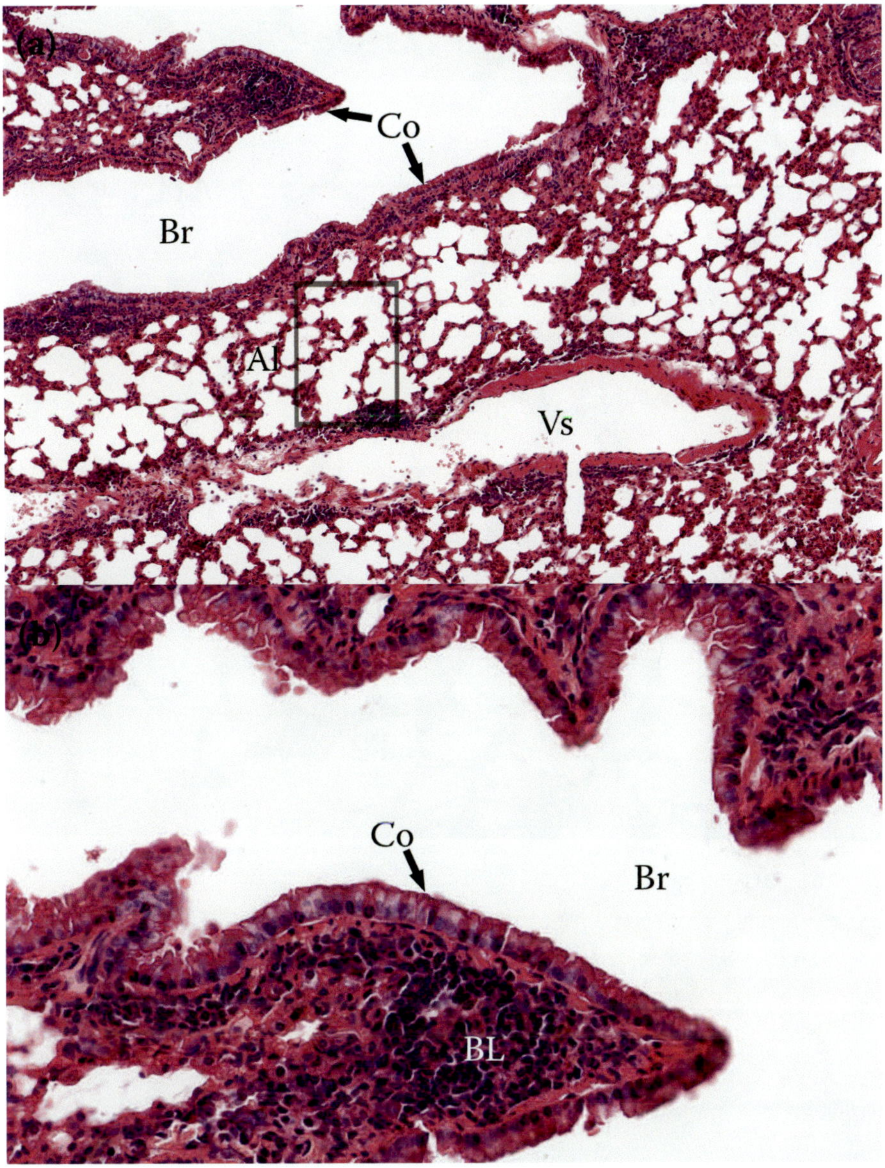

simple columnar epithelium (Co). In mouse there is no cartilage in the respiratory tree beyond primary bronchi. Epithelia of bronchioles are simple columnar. (b) Bronchus associated lymphoid tissue (BALT, H&E HP). High-power view to show simple ciliated columnar epithelium of bronchiole (Co) and bronchus associated lymphoid tissue (BL). BALT is composed of nonencapsulated lymphoid aggregates or follicles located mostly between a bronchus (Br) and artery.

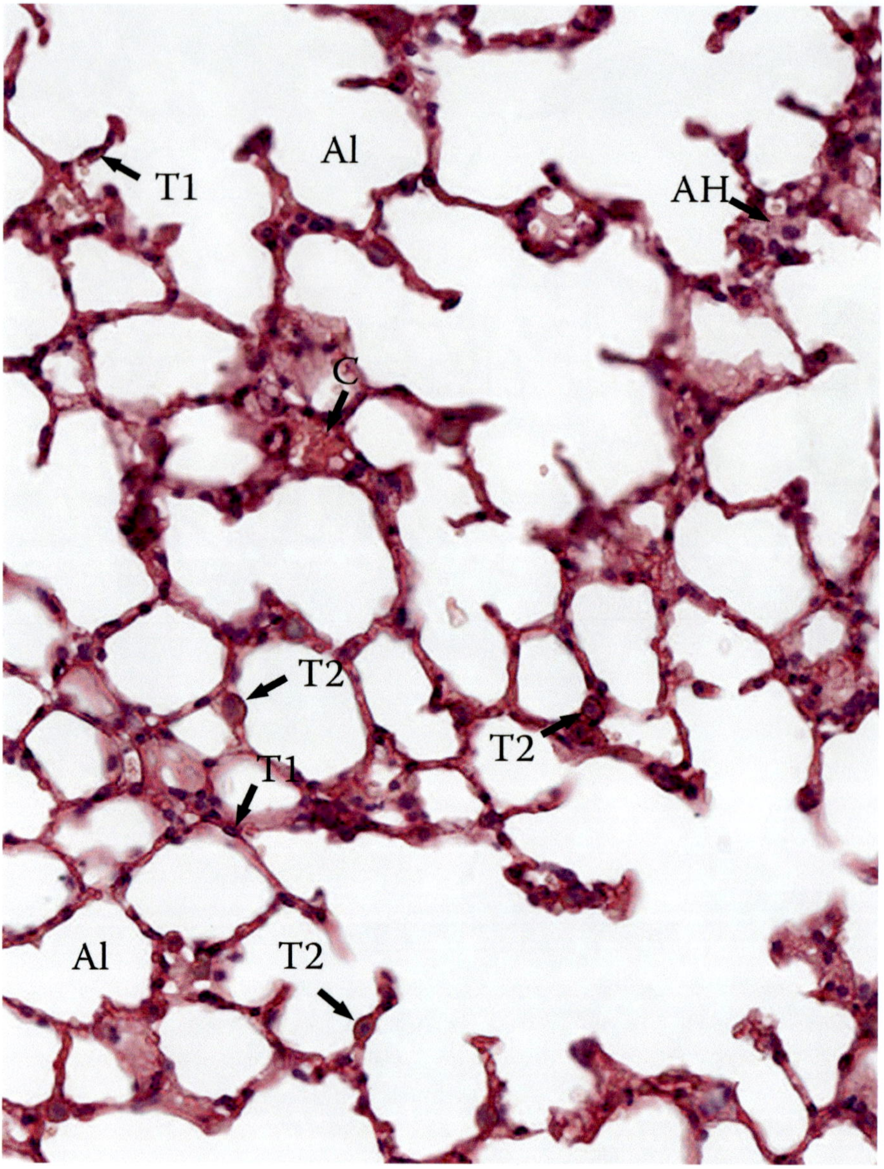

INOCULUM AND ROUTE OF INOCULATION

BALB/c Mice

The Institutional Animal Ethics Committee, registered with the Government of India (registration no. CPCSEA 1999/5), approved all animal experimental protocols and animal use. Six- to eight-week-old mice purchased from Raj Biotech, Pune, India, were randomly assigned to cages,

FIGURE 10.2 (opposite page) Normal histology (H&E, MP). A medium-power image to demonstrate Type I (T1) and Type II (T2) pneumocytes. The alveoli comprise two types of epithelial cells. Squamous epithelial or Type I cells (T1) line over 90% of alveolar surface by virtue of extremely long cytoplasmic extensions. The other main alveolar lining cell is the granular pneumocyte or Type II cell, covering about 5% of the alveolar surface (T2). They are cuboidal and by ultrastructural examination show numerous cellular organelles, including lamellar bodies, and no lateral extensions. These cells secrete surfactant. The interalveolar septa are composed of a capillary network (C) with blood separated from air by a thin layer of tissue. This tissue, sometimes referred to as the air–blood barrier, is composed of three layers: capillary endothelium, pulmonary epithelium, and intervening basement membrane. Alveolar histiocytes (AH) are resident macrophages, which play an important role in normal lung defenses against pathogens and foreign materials. They can be found in alveolar spaces (Al) or pushing between Type I pneumocytes.

with the restriction that the weights of all cage members were within 1 to 2 g of each other. They were allowed 1 week of acclimation before intake into experiments. Feed and water were given *ad libitum*.

Bacterial Strains and Inocula Preparation

Mycobacterium tuberculosis (MTB) H37Rv: The inocula were prepared by methods reported earlier (Jayaram et al., 2003). The inocula used for all the experiments were derived from a single seed lot that has been maintained at –70°C and that was made from infected mouse lungs, followed by a single round of amplification in broth. Briefly, MTB H37Rv ATCC 27294, a strain sensitive to all standard anti-TB drugs, was grown in roller bottles in Middlebrook 7H9 broth supplemented with 0.2% glycerol, 0.25% Tween 80 (Sigma), and 10% albumin dextrose catalase (Difco Laboratories, Detroit, MI) at 37°C for 7 to 10 days. The cells were harvested by centrifugation, washed twice in 7H9 broth, and resuspended in fresh 7H9 broth. Aliquots of 0.5 ml were dispensed, and the seed-lot suspensions were stored at –70°C.

Infection Procedure

Mice were infected in Biosafety Level 3 facilities via the inhalation route in an aerosol infection chamber from the Mechanical Engineering Shop, University of Wisconsin, Madison, WI (Jayaram et al., 2003). In preliminary experiments, the concentration of bacilli in a Collison nebulizer required to yield 100 CFU/lung 24 hours after infection was determined. After infection, the animals were housed for the duration of the study in Biosafety Level 3 facilities.

Histopathological Evaluation of the Lungs

Lungs from mice were collected at 8 weeks and 20 to 21 weeks postinfection and washed in PBS. They were fixed by infusing 0.3 ml of 10% buffered neutral formalin, after which they were shifted to the pathology laboratory for trimming to include gross lesions. The left lung lobes were trimmed longitudinally, and the right lung lobes were trimmed transversally. These were then processed by standard histopathological techniques using different gradation of alcohol and embedded in paraffin blocks. These paraffin-embedded tissues were sectioned at 5-µm thickness and stained with either hematoxylin and eosin (H&E) stain, Ziehl-Neelson (ZN) staining for acid-fast bacilli, or Masson's trichrome staining for demonstrating connective tissue.

FEATURES

MTB H37Rv grows logarithmically in the lungs of BALB/c mice for the first 3 weeks of infection, following an initial implantation of approximately 100 CFU per animal. By 4 weeks the bacterial load in the lungs reaches approximately 5×10^5 CFU/animal. Beyond 4 weeks postinfection, the bacterial growth slows down, and the load remains more or less constant during the remaining life of the animal. However, during the same period beyond 4 weeks postinfection, the pathology in the lung gets more complex (North, 1995; Kelly et al., 1996; Gonzalez-Juarrero et al., 2001; Munoz-Elias et al., 2005).

Gross Pathology

Grossly visible lesions appear on the surface of lung by 4 weeks. They grow from pinpoint to 2 mm by 4 weeks. By 8 weeks, consolidations with multiple pale and circular granulomatous lesions of 2- to 3-mm diameter appear in the lungs. By 20 weeks, these lesions grow slightly bigger (5 mm) and rise above the surface of the lung (Figure 10.3).

Histopathology

Microscopically, upon infection the bacilli are phagocytosed by resident macrophages in the lung called alveolar histiocytes. The host exhibits

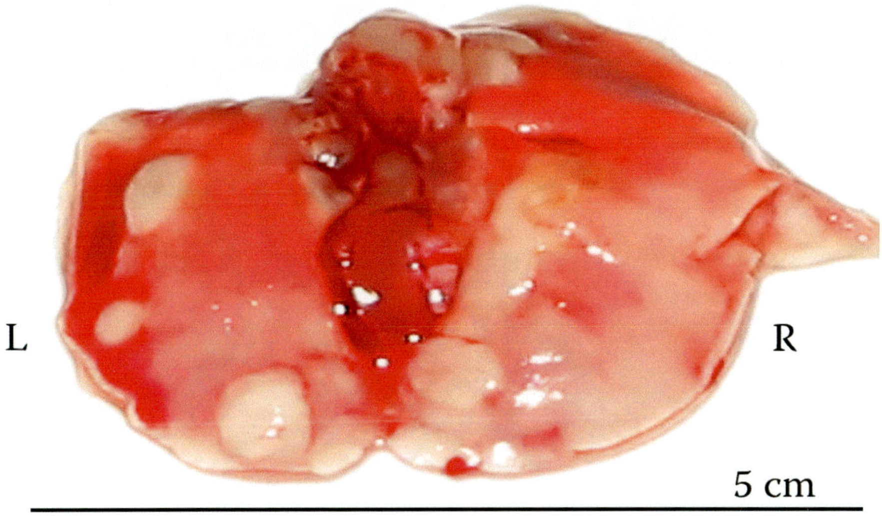

FIGURE 10.3 Gross lung at 20 weeks postinfection. BALB/c mice were infected with MTB H37Rv (ATCC27294) by the inhalation route (approximately 100 CFU/animal). This macroscopic photograph demonstrates multiple pale circular lesions (2 to 5 mm in diameter), which are visible on the pleural surface of the lung and are slightly raised.

cell-mediated immune response to infection by 3 weeks. Basic granulomatous lesions are formed by alveolar histiocytes. As the infection progresses, granulomatous bronchopneumonia of the lungs is observed.

Early Chronic Phase (8 Weeks Postinfection)

One longitudinal section of left lung of mouse shows 3 to 4 granulomatous lesions (Figure 10.4). These lesions are medium sized. Normal architecture of lung is destroyed by accumulation of infiltrated cells. Airway and alveolar spaces are infiltrated with large numbers of mononuclear cells. The lesion is predominantly cellular. The cells include epithelioid histiocytes, lymphocytes, and foam cells. Occasionally, a few degenerating neutrophils are seen. Epithelioid histiocytes show pink (eosinophilic) large cytoplasm and open nuclei but indistinct cell membranes. The cell membranes interdigitate extensively with neighboring cells of the same kind, resulting in an indistinct cell outline. The epithelioid cells are arranged in follicular pattern. Some are binucleate. In chronic lesions differentiation to epithelioid cells occurred in macrophages (Figure 10.5 and Figure 10.6).

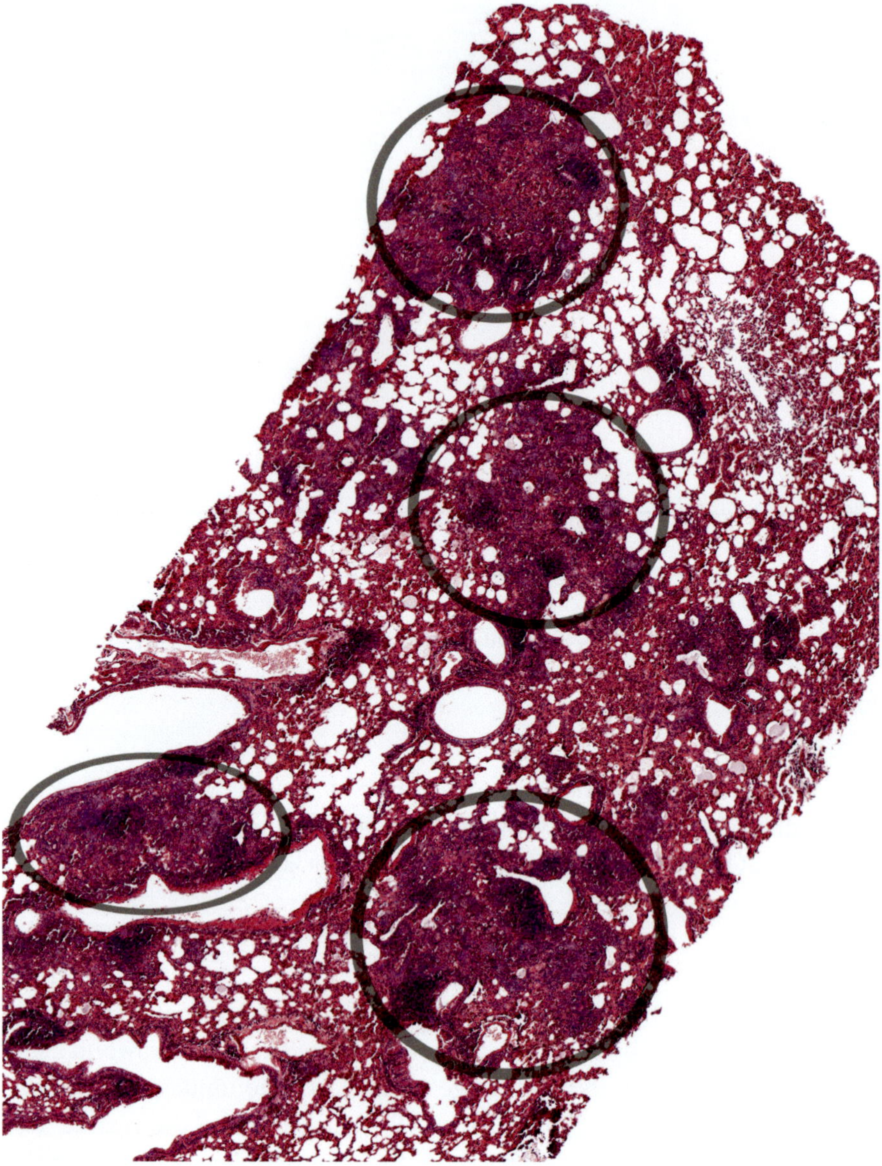

FIGURE 10.4 Tuberculous granulomas at 8 weeks postinfection (H&E, LP). The figure shows one longitudinal section of lung. Multiple granulomatous cellular lesions are present close to the conducting airways (circled) surrounded by intra-alveolar foamy macrophages.

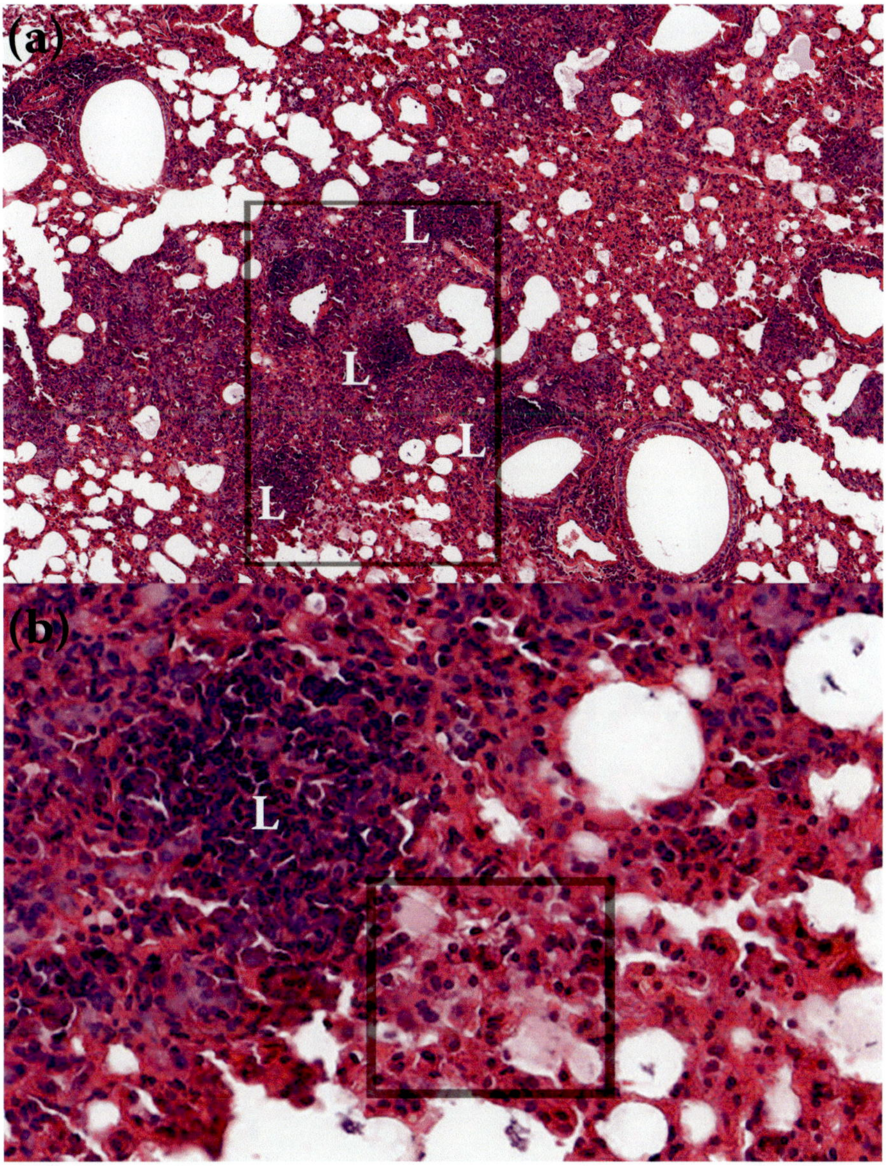

FIGURE 10.5 Granuloma with lymphoid aggregates (8 weeks postinfection, H&E, MP, and HP). (a) MP. Airway and alveolar spaces are infiltrated by ill-defined granulomatous lesions of mononuclear cells. Epithelioid histiocytes surround the aggregates of lymphocytes (L). The high power (b) demonstrates intra-alveolar foamy macrophages (square) close to the lymphocyte aggregates (L).

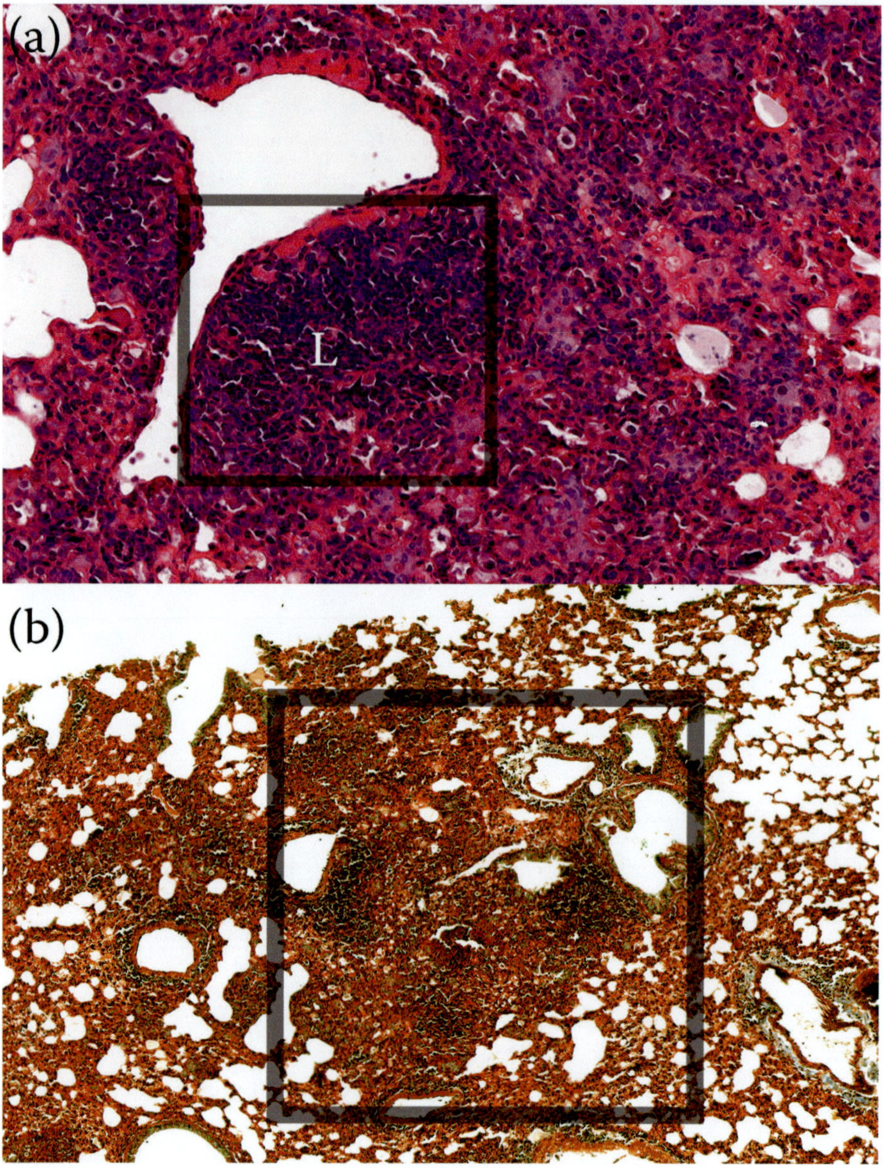

FIGURE 10.6 Eight weeks postinfection (H&E and Masson's trichrome, MP and LP). (a) Airway and alveolar spaces are infiltrated with large numbers of mononuclear cells comprising lymphocyte aggregates (square) with surrounding epithelioid histiocytes and scattered foamy macrophages. The Masson's trichrome stain at low power (b) demonstrates the spreading inflammation (square) with no evidence of connective tissue proliferation/fibrosis in the granulomatous area.

Late Chronic Phase (20 to 21 Weeks Postinfection)

Grossly, consolidations with multiple pale and raised circular granulomatous lesions of 2- to 5-mm diameter were seen (Figure 10.3). Microscopically, these lesions are focally extensive and occupy both edges of visceral pleura (i.e., mediastinal end to costal end) (Figure 10.7). In some cases the granuloma has spread to half of the lung lobe from apex. The lesions contain foam cells, lymphocytes, and epithelioid histiocytes. These cells fill the alveolar spaces, perivascular and peribronchiolar regions and are densely organized with focal degeneration. Cellular debris is seen wherever cells are undergoing degeneration (Figure 10.8). Interspersed with this cellular accumulation are cholesterol crystals, with a characteristic clefts-like appearance and a tendency to occur in picket fence-type groups (Thomson, 1984). A small focus of ossification is noticed (Figure 10.9 and Figure 10.10), which is otherwise rare in a mouse lesion (Basaraba, 2008). Scattered giant cells were also present within the granuloma (Figure 10.11). These were not typical Langhans' type giant cells.

VARIATIONS

Necrotic Granuloma

Centrally necrotic granulomas, which are a hallmark of human tuberculosis, are not observed in BALB/c mice using the aerosol tuberculosis infection model. Occasionally, cells undergo degeneration and necrosis at 20 to 21 weeks postinfection. However, this feature is not similar to the necrosis seen in human lung tuberculosis. In some genetically modified mice there are reports of neutrophilic accumulation in the granuloma (Cooper et al., 2000a,b) and also necrotic exudative inflammation in $CD4^{+}$ T cell-depleted mice (Dunn and North, 1995). Some reports state that the central necrotic zone is absent in all mouse MTB lesions even after one year of infection as demonstrated by pimonidazole adducts studies (Aly et al., 2006).

Calcified Lesions

Calcification refers to the deposition of calcium salts in soft tissues. Highly susceptible mouse strains do not exhibit calcification. This is could be due to their shortened life span secondary to rapidly progressing infection (Basaraba, 2008).

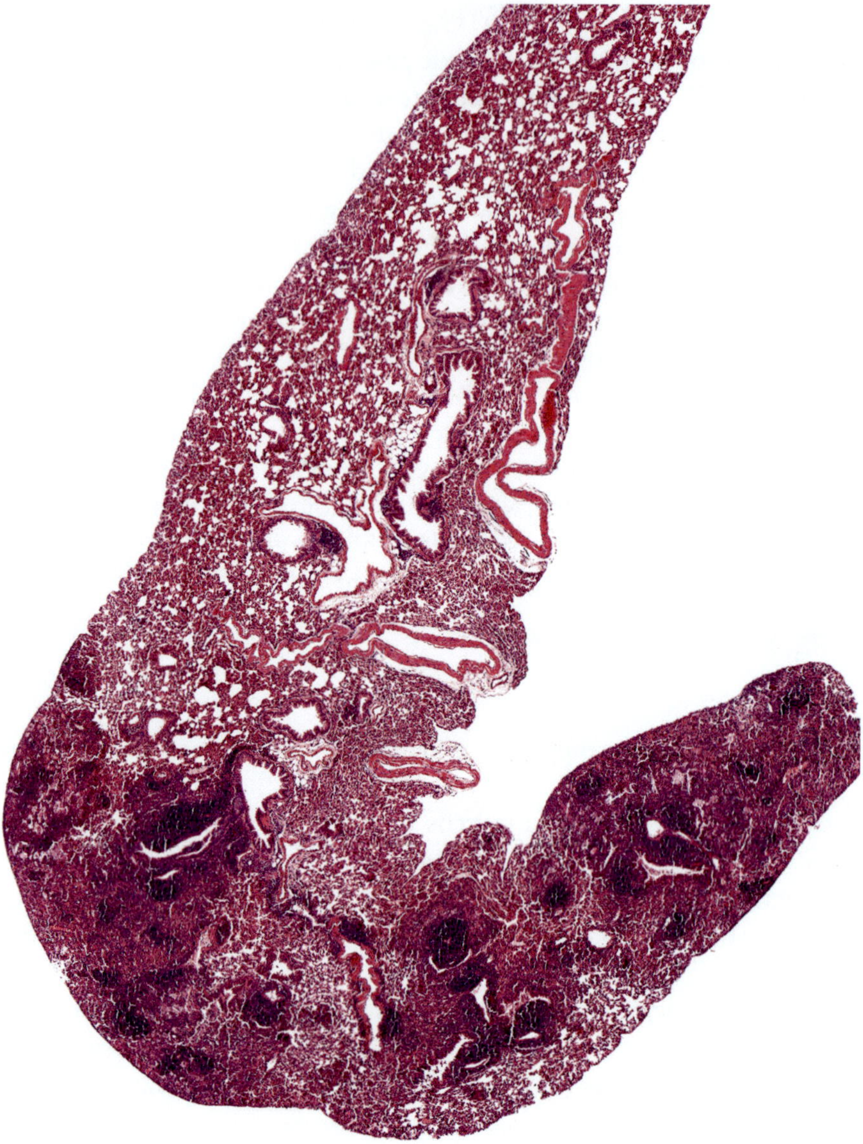

FIGURE 10.7 Twenty weeks postinfection (H&E, LP). BALB/c mice were infected with MTB H37Rv (ATCC27294) by the inhalation route (approximately 100 CFU/animal). The figure shows a low-power view of a longitudinal section from left lung at week 20 to 21 postinfection. It demonstrates focally extensive granulomas in the lower half of the section extending from the mediastinal end to costal end of visceral pleura. In contrast to the dense circular granulomatous lesions of 2-to 5-mm diameter, the upper half of the lung appears to be normal with some hyperplasia of the BALT.

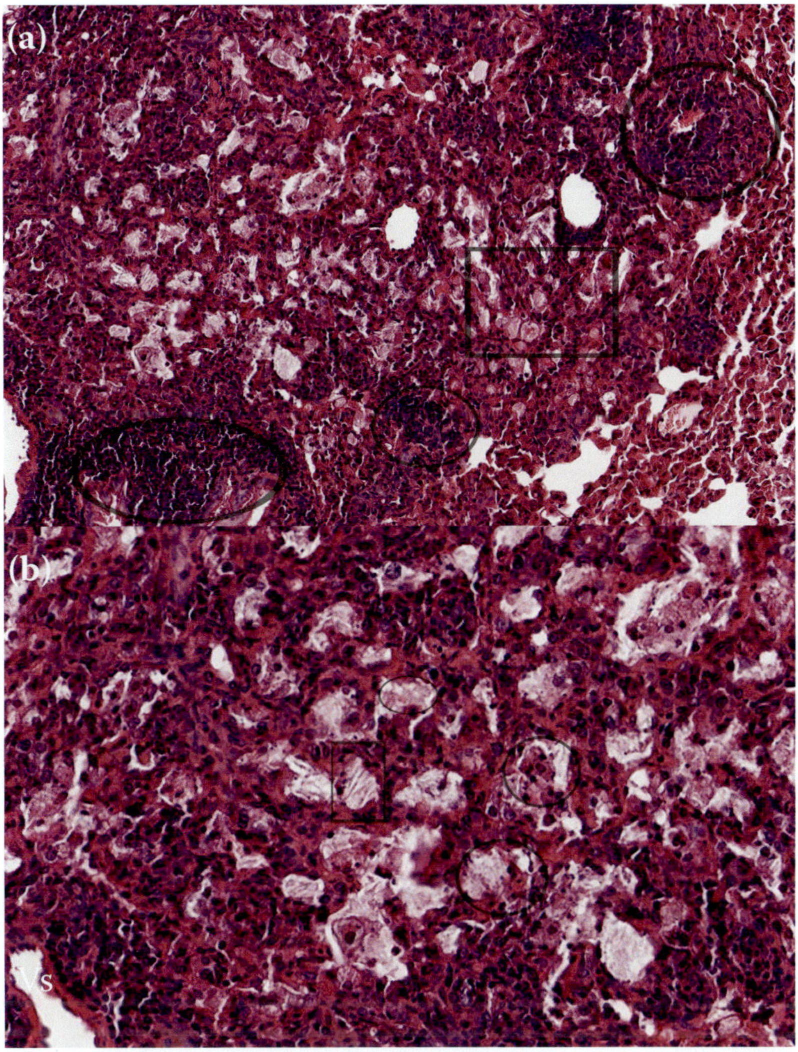

FIGURE 10.8 Granulomatous lesions, 21 weeks postinfection (H&E, MP, and HP). (a) Granulomatous lesions are focally extensive and comprise foamy macrophages (rectangle) and lymphocytes (circles) and epithelioid histiocytes. These cells fill the alveolar spaces, perivascular and peribronchiolar regions and are densely organized with focal degeneration. Large foamy macrophages with highly vacuolated cytoplasm, an important feature of tuberculous granuloma, are seen in clusters (rectangle). (b) Cellular debris is seen within the granuloma wherever cells are undergoing degeneration (circles). Focal cholesterol clefts (rectangle), a histological term to describe the space left by cholesterol crystals which are dissolved during tissue processing, are also present. These can occur in picket fence-like groups, not seen in this figure. A vascular space is present in the lower left corner (Vs).

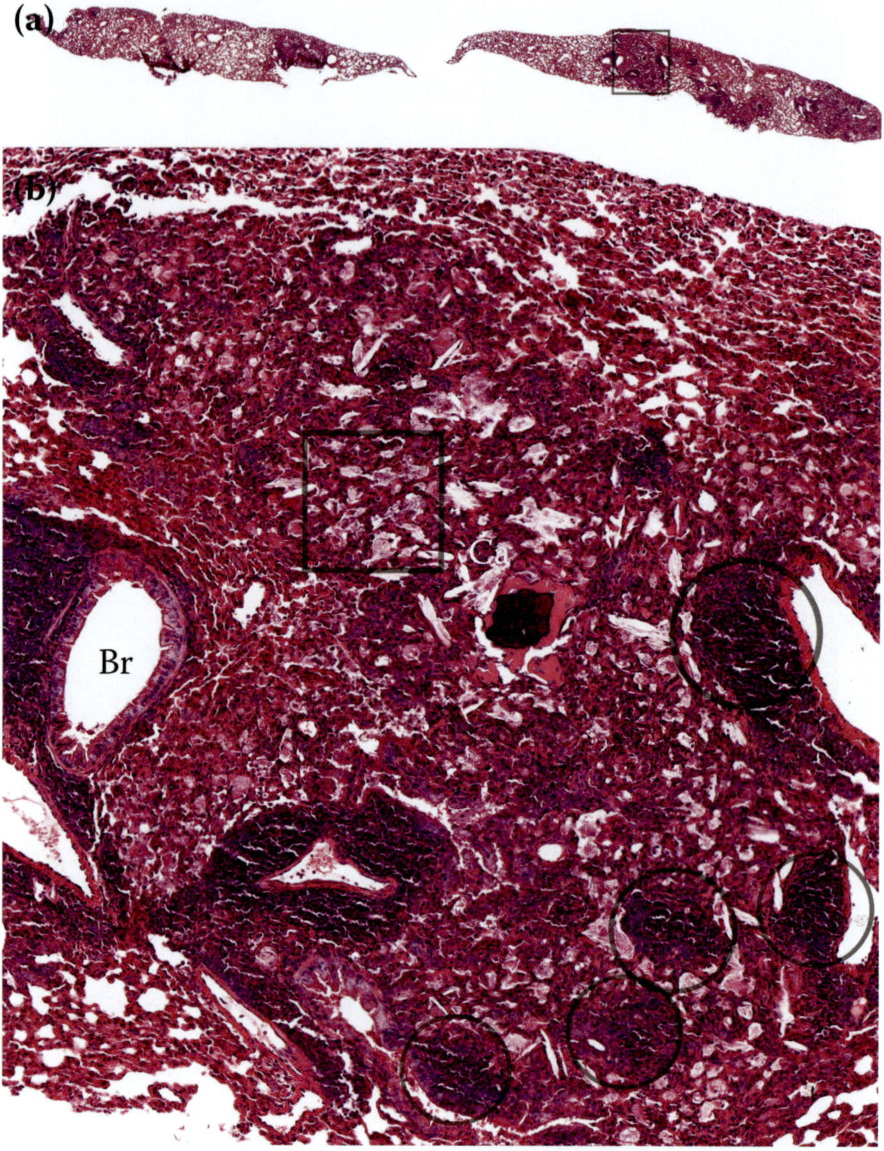

FIGURE 10.9 Ossification within a granulomatous area, 21 weeks postinfection (H&E, LP, and MP). (a) Low-power view of two longitudinal sections of lung with at least three granulomatous lesions which extend to both edges of visceral pleura. One of these areas (rectangle) is enlarged in (b). (b) The cellular infiltrate comprises lymphocyte aggregates (circles) as well as intra-alveolar epithelioid histiocytes and foamy macrophages (square). There is an area of ossification (Ca) present, generally rare in a mouse lesion and probably not associated with TB. Br—bronchiole.

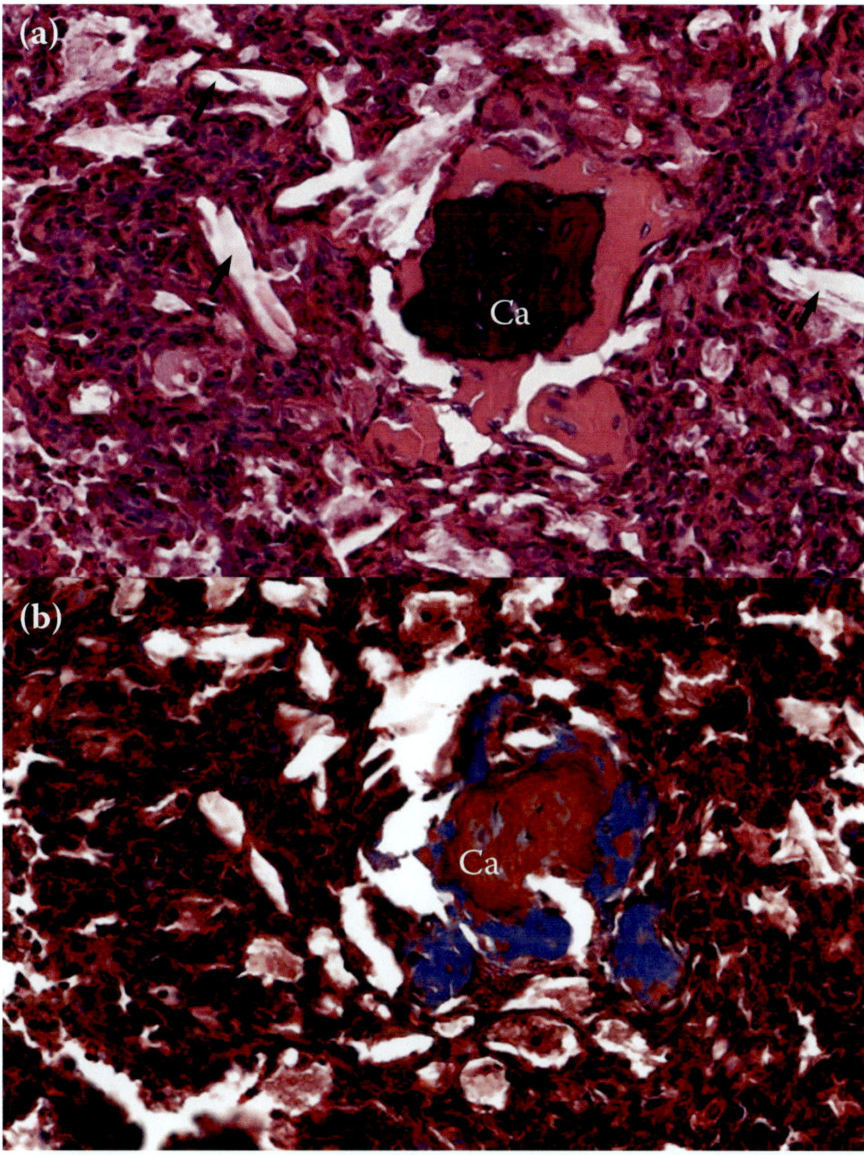

FIGURE 10.10 Ossification and cholesterol clefts, 21 weeks postinfection (H&E and Masson's trichrome, HP). (a) Higher magnification of previous figure. Ossification is uncommon in typical mouse studies. The two different stains demonstrate the need to correlate morphology with histochemistry. (b) The Masson's trichrome stains collagen blue, but can also stain foci of ossification a similar color; the central calcified area remaining eosinophilic. There is no significant fibrosis in this granulomatous area. Interspersed within this cellular accumulation are cholesterol clefts (a, arrows).

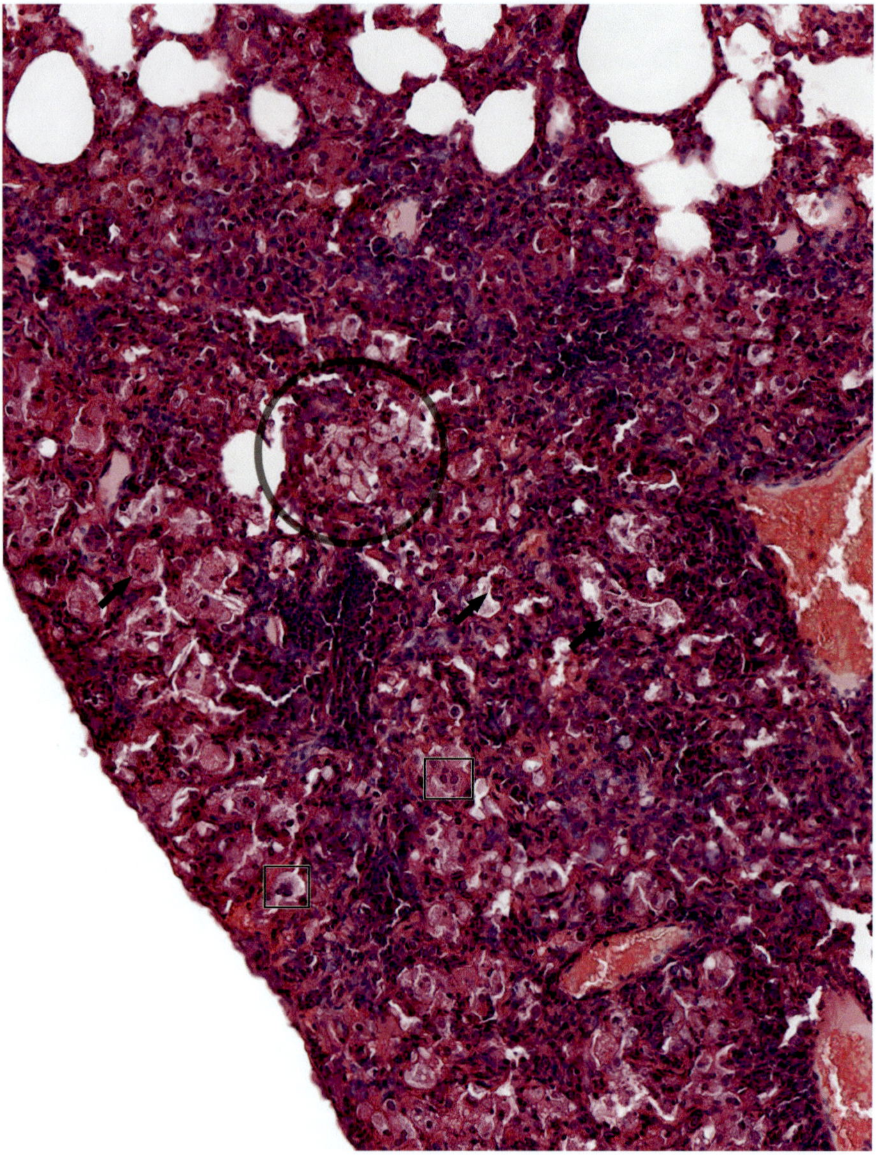

FIGURE 10.11 Spreading granulomatous inflammation with foam and giant cells, 20 weeks postinfection (H&E). This demonstrates a poorly contained spreading granulomatous inflammation comprising epithelioid histiocytes, foamy macrophages, and lymphocytes. Unlike granulomas, the process is diffuse and not discrete. Cellularity is dense and exhibits focal degeneration evidenced by cellular debris seen within the granulomatous process (arrows). Large foamy macrophages with highly vacuolated cytoplasm are seen in clusters (circle). Occasional multinucleated giant cells are visible (rectangles).

Cavitary Lesions

Human granulomas are centrally necrotic, peripherally fibrotic, and calcify or evolve into cavities, usually when bronchi are affected too. Human tuberculous lesions are circumscribed. Sections of cavitary lesions in humans reveal a layer of deeply eosinophilic fibrinous material overlying a layer of granulation tissue and densely consolidated and congested region of granuloma. There is extensive tissue destruction because of necrosis (caseous) and infiltration with Langhans' type giant cells, degenerating foamy macrophages, no epithelioid cells, little fibrosis, and few lymphocytes (Hunter et al., 2007). Mouse tubercle granulomas do not demonstrate this spectrum of lesions.

STRENGTHS OF THIS ANIMAL MODEL

Given the ease of manipulation and housing, availability of inbred strains, mutants, and genetically altered strains, and the wide range of available reagents, the mouse is by far the most practical model for tuberculosis research and for infectious disease research in general. This has led to the development and commercialization of a large number of immune reagents, thus facilitating studies of tuberculosis immunopathology during both the acute and chronic phases of the disease, within well-defined immune competent or deficient backgrounds. Mice can be infected with MTB by a variety of routes: intratracheal, intravenous, and inhalation, of which the latter is the most relevant route of transmission (Flynn, 2006). Since mice are well suited for pharmacokinetic studies with existing and experimental drugs, the mouse model of tuberculosis infection has been used to investigate pharmacokinetic/pharmacodynamic (PK/PD) relationships in several elegant studies (Shandil et al., 2007).

Implications in Drug Discovery

Given their small size and low body weight, mice are the preferred species in drug discovery due to their cost-effectiveness in terms of experimental compound requirements. Infection with MTB results in a granulomatous response, though predominantly lymphocytic in nature. This permits the investigation of PK/PD relationships of investigational drugs with reasonably predictive power for their utility in humans (Doggrell, 2005; Nikonenko et al., 2008; Nuermberger, 2008). Yet, one should keep in mind that the complex structure and diversity of granulomas and lesions found in humans likely create a barrier to effective drug penetration to the site of infection, a barrier which is not reproduced in the mouse.

SHORTCOMINGS

The two major limitations of the mouse model are (1) the absence of tuberculous granulomas that bear close resemblance to a "typical" human TB granuloma, and (2) the rapid disease progression leading to a chronic phase during which high bacillary loads are maintained for months. Many early investigators noted that human tuberculous granulomas are rich in lipids, most of which are of host origin. In contrast, mice predominantly exhibit mononuclear cell-based granulomas, and most investigations rely upon the use of single or specific inbred strains of mice. It is highly unlikely that any single strain fully captures the wide spectrum of pathophysiology and multiplicity of microenvironments that tubercle bacilli encounter in the human host (Flynn, 2006).

REFERENCES

Aly, S., Wagner, K., Keller, C., Malm, S., Malzan, A., Brandau, S., Bange, F.C., and Ehlers, S. 2006. Oxygen status of lung granulomas in *Mycobacterium tuberculosis*-infected mice. *J Pathol,* 210, 298–305.

Apt, A., and Kramnik, I. 2009. Man and mouse TB: Contradictions and solutions. *Tuberculosis (Edinb),* 89, 195–198.

Basaraba, R.J. 2008. Experimental tuberculosis: The role of comparative pathology in the discovery of improved tuberculosis treatment strategies. *Tuberculosis (Edinb),* 88 Suppl. 1, S35–47.

Cooper, A.M., Pearl, J.E., Brooks, J.V., Ehlers, S., and Orme, I.M. 2000a. Expression of the nitric oxide synthase 2 gene is not essential for early control of *Mycobacterium tuberculosis* in the murine lung. *Infect Immun,* 68, 6879–6882.

Cooper, A.M., Segal, B.H., Frank, A.A., Holland, S.M., and Orme, I.M. 2000b. Transient loss of resistance to pulmonary tuberculosis in p47(phox–/–) mice. *Infect Immun,* 68, 1231–1234.

Dharmadhikari, A.S., and Nardell, E.A. 2008. What animal models teach humans about tuberculosis. *Am J Respir Cell Mol Biol,* 39, 503–508.

Doggrell, S.A. 2005. New drugs being developed for the treatment of tuberculosis. *Expert Opin Investig Drugs,* 14, 917–920.

Dunn, P.L., and North, R.J. 1995. Virulence ranking of some *Mycobacterium tuberculosis* and *Mycobacterium bovis* strains according to their ability to multiply in the lungs, induce lung pathology, and cause mortality in mice. *Infect Immun,* 63, 3428–3437.

Flynn, J.L. 2006. Lessons from experimental *Mycobacterium tuberculosis* infections. *Microbes Infect,* 8, 1179–1188.

Gonzalez-Juarrero, M., Turner, O.C., Turner, J., Marietta, P., Brooks, J.V., and Orme, I.M. 2001. Temporal and spatial arrangement of lymphocytes within lung granulomas induced by aerosol infection with *Mycobacterium tuberculosis*. *Infect Immun,* 69, 1722–1728.

Hunter, R.L., Jagannath, C., and Actor, J.K. 2007. Pathology of postprimary tuberculosis in humans and mice: Contradiction of long-held beliefs. *Tuberculosis (Edinb),* 87, 267–278.

Jayaram, R., Gaonkar, S., Kaur, P., Suresh, B.L., Mahesh, B.N., Jayashree, R., Nandi, V., Bharat, S., Shandil, R.K., Kantharaj, E., and Balasubramanian, V. 2003. Pharmacokinetics–pharmacodynamics of rifampin in an aerosol infection model of tuberculosis. *Antimicrob Agents Chemother,* 47, 2118–2124.

Kelly, B.P., Furney, S.K., Jessen, M.T., and Orme, I.M. 1996. Low-dose aerosol infection model for testing drugs for efficacy against *Mycobacterium tuberculosis*. *Antimicrob Agents Chemother,* 40, 2809–2812.

McMurray, D.N., Collins, F.M., Dannenberg, A.M., Jr., and Smith, D.W. 1996. Pathogenesis of experimental tuberculosis in animal models. *Curr Top Microbiol Immunol,* 215, 157–179.

Munoz-Elias, E.J., Timm, J., Botha, T., Chan, W.T., Gomez, J.E., and McKinney, J.D. 2005. Replication dynamics of *Mycobacterium tuberculosis* in chronically infected mice. *Infect Immun,* 73, 546–551.

Nikonenko, B.V., Sacksteder, K.A., Hundert, S., Einck, L., and Nacy, C.A. 2008. Preclinical study of new TB drugs and drug combinations in mouse models. *Recent Pat Antiinfect Drug Discov,* 3, 102–116.

North, R.J. 1995. *Mycobacterium tuberculosis* is strikingly more virulent for mice when given via the respiratory than via the intravenous route. *J Infect Dis,* 172, 1550–1553.

Nuermberger, E. 2008. Using animal models to develop new treatments for tuberculosis. *Semin Respir Crit Care Med,* 29, 542–551.

Shandil, R.K., Jayaram, R., Kaur, P., Gaonkar, S., Suresh, B.L., Mahesh, B.N., Jayashree, R., Nandi, V., Bharath, S., and Balasubramanian, V. 2007. Moxifloxacin, ofloxacin, sparfloxacin, and ciprofloxacin against *Mycobacterium tuberculosis*: Evaluation of in vitro and pharmacodynamic indices that best predict in vivo efficacy. *Antimicrob Agents Chemother,* 51, 576–582.

Thomson, R.G. 1984. *General veterinary pathology,* 2nd ed. Philadelphia: W.B. Saunders Company.

Glossary

Acid-fast: Bacteria that are not decolorized by an acidic alcohol solution after they have been stained. See also *Ziehl-Neelsen stain.*

Acute-phase: In the context of TB animal models only, acute phase refers to the period immediately after infection during which the number of bacilli is increasing. As we have no means of measuring bacilli numbers after primary infection in man, there is no human correlate. See also *chronic phase.*

Adipose: Referring to fat cells (adipocytes) or fat.

Anthracosis: A general term for describing carbon dust deposition in the lung and lymph nodes, which in isolation does not cause disease. It is usually present in urban dwellers, and in those working in certain occupations such as coal mining. Anthracosis can also refer to a chronic lung disease characterized by the deposit of coal dust in the lungs and by the formation of black nodules on the bronchioles, resulting in focal emphysema. The condition occurs in coal miners, is aggravated by cigarette smoking, and is also called black lung disease, coal worker's pneumoconiosis, and miner's pneumoconiosis.

Basal lamina: A layer of extracellular matrix (typically 40 to 50 nm thick, though thicker in the glomerular basement membrane) secreted by epithelial cells on which epithelium sits. Often confused with the basement membrane, "basal lamina" is usually used with electron microscopy, while the term "basement membrane" is usually used with light microscopy. Basement membrane in light microscopy refers to the stained structure anchoring an epithelial layer. This encompasses the basal lamina and typically a reticular lamina secreted by other cells.

Bronchocentric: Surrounding a bronchus.

Capsule: A structure enclosing an organ or an area of identifiable histological change, usually composed of dense connective tissue.

Caseation/caseous necrosis: The characteristic central area of tissue necrosis seen in tuberculous lesions, most evident in post-primary lesions. The term caseous necrosis or caseation refers to the cream (cottage) cheese-like appearance. This is a macroscopic term and should not be used for histological description. The histological term would be "necrotizing."

Cavitation, pulmonary: In the context of TB, this refers to a necrotic tuberculoma communicating with an airway, enabling TB bacilli to enter the sputum and to be coughed out.

Cell-mediated immunity, cellular immunity: The mechanism of acquired immunity characterized by the dominant role of T cell lymphocytes. Cell-mediated immunity is key to resisting infectious diseases caused by viruses and some bacteria, involved in delayed hypersensitivity reactions, some aspects of cancer resistance, certain autoimmune diseases, graft rejection, and certain allergies. It does not involve the production of humoral antibody, but instead involves the activation of monocytes/macrophages and natural killer cells. See also *humoral immunity.*

Cholesterol cleft: An artifactual space caused by the dissolving of cholesterol crystals in sections of paraffin-embedded tissue.

Chronic inflammatory cells: The cells which play a role in chronic inflammation, namely all the inflammatory cells except for neutrophils. This includes lymphocytes, plasma cells, mononuclear cells (monocytes, macrophages), eosinophils, basophils, and mast cells.

Chronic phase: In the context of animal models only, chronic phase refers to the period after infection during which bacilli numbers are stable and not increasing. It occurs after acute phase and can be divided into early and late periods. See also *acute phase.*

Congestion: In the context of lung disease, pulmonary congestion refers to dilatation of vasculature often secondary to left heart dysfunction and consequent back pressure. Dilated capillaries are filled with erythrocytes, and the increased hydrostatic pressure results in transudation of plasma fluid into the alveolar spaces.

Consolidated: A macroscopic description referring to lung which has become solid and firm in appearance, in contrast to normal spongy, elastic, air-filled spaces. Consolidation is due to accumulated fluid and inflammatory debris and occurs in pneumonia, which may be due to various infective causes, TB being one of them.

Cryptic disseminated tuberculosis: Widespread tuberculosis occurring in the immunosuppressed, such as those with AIDS. The tissues contain numerous microscopic lesions with abundant TB bacilli. In contrast to the lesions of miliary tuberculosis, these are not easily seen on radiology, hence the term cryptic.

Disseminated: Spread.

Disseminated tuberculosis: Tuberculosis that has spread from the original site of infection to involve multiple organs and tissues. There are two types—cryptic disseminated and miliary.

Dyspnea: Shortness of breath.

Dystrophic: Occurring at sites of damage or necrotic tissue, often as a reaction to tissue damage. Dystrophic calcification is calcification occurring in degenerate or necrotic tissue, as in hyalinized scars, degenerated foci in leiomyomas, and necrotizing granulomas.

Emphysema: This is a pathologic term (rather than a specific disease), which is defined as the abnormal permanent enlargement of airspaces distal to the terminal bronchioles, accompanied by destruction of their walls, without obvious fibrosis. It can be subdivided into centrilobular emphysema (associated with smoking, and affecting the lobules around the central respiratory bronchioles in the upper lung zones) and panlobular emphysema (associated with alpha-1-antitrypsin deficiency, and uniformly affecting the entire secondary lobule, commonly in the lower lung zones). Emphysematous changes result in loss of lung elasticity and consequent collapse during exhalation. Hyperinflation and air trapping in emphysematous lung causes external airway compression and obstruction. Emphysema is part of the spectrum of chronic obstructive pulmonary disease (COPD).

Epithelioid: Having the appearances of an epithelial cell, i.e., relatively large (4× to 5× RBC area), low nuclear:cytoplasmic ratio, pale vesicular nucleus, abundant cytoplasm. Epithelioid histiocytes are the major component of granulomas.

Exogenous reinfection of tuberculosis: Tuberculosis which occurs in a person with a previous TB infection. This represents a recent reinfection.

Extensively Drug-Resistant TB (XDR-TB): The WHO Global Task Force on XDR-TB in October 2006 defined XDR-TB as TB that has developed resistance to rifampicin and isoniazid (MDR-TB), as well as to any member of the quinolone family and at least one of the following second-line anti-TB injectable drugs: kanamycin, capreomycin, or amikacin. Earlier definitions of XDR-TB may refer to it as MDR-TB that is also resistant to three or more of the six classes of second-line drugs.

Extrapulmonary tuberculosis: Tuberculosis occurring outside the lung parenchyma (tissue), but including lesions within the thoracic cage. For example, a TB granuloma found in the pleural on the thoracic wall would be extrapulmonary but intrathoracic.

Extrathoracic tuberculosis: TB occurring outside the thoracic cage.

Fibroblast: A cell which forms part of the connective tissue framework in animal tissues, synthesizes extracellular matrix and collagen, and plays a role in wound healing. They are morphologically heterogeneous with diverse appearances depending on their location and activity and can appear as thin spindly cells through to plump cells with multiple nucleoli and branching cytoplasmic projections. They will be active around a TB granuloma.

Fibrosis: A proliferation of fibrous connective tissue that occurs normally in the formation of scar tissue to replace tissue lost through injury or infection. Fibrosis is most common in the heart, lung, peritoneum, and kidney.

Foam cells, foamy macrophages, foamy histiocytes: Foam cells are lipid-laden macrophages. They are easily recognized by their abundant cytoplasm, which appears "bubbly." These macrophages are common in areas of inflammation and cell injury and acquire this appearance through phagocytosis of lipid from injured cells.

Ghon focus: The initial focus of tuberculous infection in the lung, named after Austrian pathologist Anton Ghon (1866–1936) who is best known for his 1912 work on childhood tuberculosis called *Der primäre Lungenherd bei der Tuberkulose der Kinder.* A Ghon focus is typically a parenchymal subpleural lesion, either just above or just below the interlobar fissure between the upper and lower lobes. The Ghon complex is a Ghon focus plus enlarged caseous lymph nodes draining the parenchymal focus.

Giant cell: A giant cell is an amalgamation of macrophages as a response to indigestible material, which may be a foreign body or from tissue breakdown. They are multinucleated and 20× to 30× larger than the average macrophage. Historically, pathologists have given their names to those seen in particular diseases or with particular appearances (e.g., Touton's giant cell, Langhans' giant cell). However, none of these giant cells have been found to be specific to the disease for which they were originally described. When in doubt, a generic description that can be used is "foreign body-type giant cell." Presence of foreign material can often be confirmed using plane-polarized light.

Granulocyte: White blood cells which contain granules in their cytoplasm. Also referred to as polymorphonuclear leukocytes because of the multilobated nature of the nucleus. Neutrophils, eosinophils, and basophils are all granulocytes.

Granuloma: A localized aggregate or collection of epithelioid histiocytes. It may display central necrosis, in which case "necrotizing granuloma" is applied. Granulomas in the TB infection (sometimes referred to as tuberculous granulomas or macroscopically as tuberculomas) usually will also demonstrate a rim of lymphocytes and scattered plasma cells external to the epithelioid cells and an outermost rim of fibroblasts or fibrosis. Langhans' type giant cells with multiple nuclei arranged in a horseshoe pattern are often present among the epithelioid cells.

Granulomatous inflammation: As distinct from granuloma which is localized, granulomatous inflammation shows a similar composition of cell types but is diffuse and not localized or ball-like.

Granulation tissue: A normal tissue response composed of proliferating fibroblasts and capillaries with variable numbers of inflammatory cells including macrophages, plasma cells, and lymphocytes. Granulation tissue represents the healing phase in the reaction to injury and tissue damage. It is preceded by necrosis and inflammation and followed by fibrosis and regeneration.

Hematoxylin and eosin staining (H&E): The most commonly used histochemical stain in diagnostic pathology. It uses the basic dye hematoxylin, which colors basophilic structures with blue-purple hue, and alcohol-based acidic eosin Y, which colors eosinophilic structures bright pink. Basophilic structures usually contain nucleic acids—ribosomes, chromatin-rich cell nucleus, and cytoplasmic regions rich in RNA. Eosinophilic structures are generally composed of intracellular or extracellular protein. Most of the cytoplasm is eosinophilic. Red blood cells stain intensely red. Structures do not have to be acidic or basic to be called basophilic and eosinophilic because the terminology is based on the affinity to the dyes. Hydrophobic structures will not take dye. Other colors can be present in an H&E stain, and these are usually caused by intrinsic pigments (e.g., melanin, carbon deposition).

Heterophil: A granular polymorphonuclear leukocyte, which in humans is represented by the neutrophil. In other mammals, the granules (lysosomes) have variable sizes and staining characteristics and may look similar to human eosinophils.

High power: See *magnification*.

Hilus or hilum: A depression in an organ where the vessels and nerves and other structures, in this case bronchi, enter or leave. The lung hilum is located centrally between the left and right lung.

Histiocyte: A histiocyte is a tissue macrophage, a type of white blood cell, that is derived from the monocyte in the bone marrow. Some histiocytes may not originate from the marrow but from pluripotential cells in the connective tissue. Histiocytes usually stay in place, but when stimulated by infection or inflammation they become active, attacking bacteria and other foreign matter in the body. A macrophage in connective tissue is referred to as a histiocyte and this is the preferable term.

Histogenesis: The formation and development of the tissues of the body. The differentiation of pluripotential stem cells into specialized tissues.

Humoral immunity: Immunity mediated by secreted antibodies (as opposed to cell-mediated immunity, which involves T lymphocytes) produced in the cells of the B lymphocyte lineage (B cell). Humoral immunity is so named because it involves substances found in the humors (body fluids).

Immunohistochemistry: Microscopic localization of specific antigens in tissues through staining with antibodies labeled with fluorescent or pigmented material.

Immunosuppressed: Having an immune system which is inadequate or below normal.

Inflammation: A protective response to tissue injury. Inflammation acts to destroy, diminish, or contain both the injured tissue and the injurious agent. It need not be caused by infection. Inflammation can be divided into two main groups—acute and chronic—and several subtypes (e.g., granulomatous, exudative, fibrinous, hyperplastic, interstitial, parenchymatous, pseudomembranous). The classical signs of acute inflammation are pain (dolor), heat (calor), redness (rubor), swelling (tumor), and loss of function (functio laesa). Histologically, acute inflammation is characterized by neutrophils, necrosis, and vascular congestion. Chronic inflammation is prolonged and persistent inflammation in which there is new connective tissue formation and mostly chronic inflammatory cells (lymphocytes, plasma cells, eosinophils). It may be a continuation of acute inflammation or a prolonged low-grade form. Granulomatous inflammation is a chronic inflammation in which there is formation of granulation tissue.

Ipsilateral: On the same side. *Contralateral* refers to the opposite side.

Langhans' multinucleated giant cells (Theodor Langhans, 1839–1915): These are large cells with many nuclei, often in a "horseshoe" formation. They are regularly seen in tuberculous lesions, but are not specific for TB and also occur in granulomas due to other conditions. They should not be confused with Langerhans' cells (Paul Langerhans 1847–1888), which are dendritic cells in the stratum spinosum layer of the epidermis containing large granules called Birbeck granules and are also found in lymph nodes and other organs. The significance of Langhans' giant cells is debated. It is mentioned in this text because, historically, this has been the giant cell type associated with TB.

Latent tuberculosis: A term applied to the status of those immunosensitized with the tubercle bacillus (as determined by tuberculin skin testing, such as the Mantoux test) but remaining healthy. This is a clinical term. In human histopathology, it is not correct to call a lesion "latent" on histopathology alone unless you know the clinical status of the patient and confirm that the lesion you are looking at is the only lesion in the body. A granuloma with a paucity of cellular activity composed mostly of sclerotic tissue and/or calcification would be described as a "sclerotic granuloma" or perhaps a "healed" lesion or even a "quiescent" granuloma. However, this term is sometimes used in animal models where all infection parameters have been controlled, and as such, it is possible to define lesions as being part of "latent" tuberculosis.

Liquefactive necrosis: See *necrosis.*

Low power: See *magnification.*

Lumen: The cavity or channel within a tube or hollow organ.

Lymphocyte: A white blood cell originating from stem cells and differentiating in lymphoid tissue. Lymphocytes form part of the immune system and can be subdivided into Natural Killer (NK) cells, T cells, and B cells. Histologically, lymphocytes are small blue cells with very little visible cytoplasm.

Lymphoplasmacytic: Comprising both lymphocytes and plasma cells. The typical TB granuloma will have lymphocytes, plasma cells, and eosinophils in the periphery. Eosinophils are in the minority and sometimes difficult to see, so many will describe the inflammatory cell infiltrate as "lymphoplasmacytic," though a longer phrase such as "lymphocytes, plasma cells, and occasional eosinophils" is more accurate.

Macrophage: One of the classes of phagocytic cells and particularly involved in chronic infections such as tuberculosis. It is more correct to use the term histiocyte in the context of tuberculous granulomas.

Macroscopic: Referring to what can be seen and described with the naked eye. Also referred to as "gross." For example, a histopathology report will have a "macroscopic" or "gross" description detailing the findings at sectioning and a microscopic description detailing the histological findings. Some terminology such as "caseation" originates from macroscopic description and does not translate into microscopic terms.

Magnification; low power, medium power, high power: Microscope objective magnifications are not standardized nor are they exactly the magnification stated. There are also differences between manufacturers. This is compounded by use of other optical adaptors between the camera and the microscope which may increase or decrease the actual magnification. A so-called "20×" image from one microscope may show cells of different size from another microscope of different manufacture. Another reason why stating objective magnification is misleading is that it provides no information of the numerical aperture of the objective lens, the true determinant of resolution. Numerical apertures will differ between different grades of objectives of the same magnification. Additionally, if digital camera resolution is not correctly configured to complement the optical resolution, your final captured image may not demonstrate the same amount of detail as is seen through the microscope eyepieces. For this reason it is better to think in terms of three main scales—low power, medium power, and high power. Low power is for appreciating architecture and tissue textures/colors. High power is for appreciating nuclei detail and cellular cytology, and medium power sits somewhere in between.

Masson's trichrome staining: A histochemical stain used to differentiate between collagen and smooth muscle. Muscle is stained red, collagen stains blue, fibrin is pink, erythrocytes are red, and nuclei are blue/black. Mineralized bone can also stain blue, and consequently so can calcific foci in a granuloma.

Medium power: See *magnification.*

Miliary tuberculosis: A form of disseminated tuberculosis occurring in patients with a relatively good immune response. In Latin, milium is a millet seed. The pattern of miliary tuberculous is characterized by multiple small TB granulomas scattered through one or more organs in a diffuse pattern due to seeding of bacteria through lymphatics and blood vessels. Lesions may be nonnecrotizing or necrotizing. Macroscopically, it may appear that small lesions have no necrosis, but this is often not the case when examined histologically. Miliary tuberculosis is a macroscopic term which describes millet-seed sized granulomas that are easily seen on chest radiographs and, sometimes, on the retina by use of an ophthalmoscope. Miliary lesions differ from those of cryptogenic disseminated tuberculosis in that they are larger and radiologically visible. Avoid using this term when describing microscopy images.

Mucosa: A mucous membrane, comprised of several layers—epithelium, lamina propria, and often a muscularis mucosae.

Multidrug-resistant TB (MDR-TB): MDR-TB is TB resistant to at least isoniazid (INH) and rifampicin (RMP). Isolates that are multiply resistant to any other combination of anti-TB drugs but not to INH and RMP are not classified as MDR-TB. MDR-TB mostly develops during treatment as a result of inappropriate treatment or poor compliance with pharmacotherapy (e.g., missing doses or failing to complete a course of treatment). Multidrug-resistant tuberculosis can be cured but requires long treatments (1 to 2 years) of "second-line" drugs, which are more expensive than first-line drugs and have more adverse effects. See also *Extensively Drug-Resistant TB (XDR-TB).*

Necrosis: Premature death of cells and living tissue due to external factors such as injury, infection, malignancy, infarction, poisons, inflammation. It can be subdivided by morphology. Some examples include:

Liquefactive necrosis, which is associated with abscess formation and commonly found in the central nervous system, is characterized by cellular digestion by hydrolytic enzymes, resulting in space filled with pus and fluid. It is characteristic of focal bacterial or fungal infections.

Coagulative necrosis, in which cell outlines remain after cell death, is typically seen in hypoxic environments such as following infarction.

Caseous necrosis, which is seen in TB, is a specific form of coagulative necrosis. It can be caused by mycobacteria and also fungi and is a combination of coagulative and liquefactive necrosis.

Necrotizing: Causing necrosis, the histological description of caseous necrosis.

Neutrophil: A type of white blood cell, a form of *granulocyte,* highly motile, found in blood and tissue, and filled with sacs of enzymes that help the cell kill and digest microorganisms it has engulfed by phagocytosis. The mature neutrophil has a segmented nucleus (two to five lobes) and a total life span of about three days. It stains neutral pink with H&E staining. They are the hallmark of the acute inflammatory response and have the properties of chemotaxis (migration toward sites of infection/inflammation), phagocytosis (internalization of microbes or foreign particles), degranulation (release of proteins), and neutrophil extracellular traps (NETs, extracellular webs of chromatin and serine proteases which create a physical barrier for trapping and killing microbes).

Numerical aperture: Numerical aperture of a microscope objective lens is the primary determinant of resolution and more significant than the magnification. Resolution is proportional to the wavelength of light used for illumination, the refractive index of the imaging medium, and the objective angular aperture. Microscope objectives are designed to image specimens either with air or a medium of higher refractive index (such as oil) between the front lens and the specimen. A gain in resolution by a factor of approximately 1.5 is attained when immersion oil is substituted for air as the imaging medium. Numerical aperture is one of the most important factors (other than optical correction) to consider when selecting a microscope objective. Values range from 0.1 for very low magnification objectives (1× to 4×) to as much as 1.6 for high-performance objectives utilizing specialized immersion oils.

Objective: The lens unit in a microscope that gathers light and focuses the light from a slide (the object) to produce an image. Often composed of multiple lenses, microscope objectives are characterized by two main parameters—magnification and numerical aperture.

Ossification: Bone tissue formation. Not to be confused with calcification, which is the formation of calcium-based salts and crystals within tissue and is a process which occurs during ossification but not vice versa.

Pathognomonic: Characteristic for a particular disease. A pathognomonic finding is a feature whose presence means that a particular disease is present without doubt. A pathognomonic sign of symptom is highly specific, but need not be sensitive, meaning that it need not be present. The presence of a pathognomonic finding implies that the diagnosis is certain.

Parenchyma: A general term used to differentiate the functional or essential elements of an organ from its framework or stroma (supporting connective tissue).

Pathogenesis: The course of a disease from its origin to initial manifestation through to critical development and outbreak.

Pharmacodynamics (PD): The study of the physiological effects of drugs on the body or on microorganisms or parasites within or on the body and the mechanisms of drug action. PK/PD relationships refer to the study of correlations between drug concentration and effect.

Pharmacokinetics (PK): The study of the bodily absorption, distribution, metabolism, and excretion of drugs. This is most often assessed by establishing the concentration–time profile of the drug in plasma and tissues, followed by computer-assisted modeling to provide pharmacokinetic parameters.

Plasma cell: A mature B cell lymphocyte, a white blood cell that produces antibodies. Plasma cells originate in the bone marrow as B cells and then differentiate in the plasma cells in the germinal centers of lymph nodes following stimulation by T cells. On light microscopy they have a basophilic cytoplasm, high nucleus-to-cytoplasmic ratio, and eccentric nucleus with heterochromatin in a characteristic clock face arrangement.

Pleura: The membranes forming a sac or cavity in which the lungs lie.

Pleural effusion: Fluid accumulating in the pleural cavity.

Pleurisy: Inflammation of the pleura. This can occur in the pleura overlying an area of tuberculous pneumonia or if a TB granuloma extends to the pleural surface.

Postprimary tuberculosis: Tuberculosis occurring after a period of latency following initial infection. This is usually several years after initial infection. The usual site, irrespective of the site of the initial infection, is the apex of the lung. Tissue necrosis leading to cavity formation is a common finding in severe cases.

Power: See *magnification.*

Primary complex, tuberculous: The initial lesion of tuberculosis, consisting of a lesion at the site of implantation of the bacilli (usually the lung but occasionally other sites) and in the lymph nodes draining that site.

Primary tuberculosis: Active tuberculosis following initial infection.

Resolution: In very simple terms, this is the ability of an imaging system to discern detail in the object being imaged. The better your resolution, the more detail you will be able to visualize. If two points are closer together than your resolution, it will not be possible to distinguish them as separate points—they will appear as a single object. As a rough guide, an optical microscope can resolve to 200 to 370 nanometers.

Sarcoid-like: Having the appearances of a sarcoid granuloma, i.e., nonnecrotizing, well-circumscribed granuloma(s). Sarcoid disease produces nonnecrotizing granulomas.

Septum (pl. septa): The dividing wall or partition usually between lobes of an organ.

Serosa: A membrane comprising mesothelium and underlying connective tissue, lining the serous cavities of the body.

Skin test: See *Tuberculin skin test.*

Stroma: The supporting tissue or scaffolding of an organ, as distinguished from its functional element, or parenchyma. Stroma may also be referred to as connective tissue and may comprise fibroblasts, adipocytes, immune cells, collagenous tissue, elastin, structural glycoproteins such as fibronectin and laminin, and ground substances such as glycosaminoglycans.

Suppurative: Forming pus, an aggregate of neutrophils. A suppurative granuloma is a necrotizing granuloma with a large collection of neutrophils (pus) with necrosis.

Surfactant: In general, surfactant is an amphiphilic (containing both hydrophobic and hydrophilic groups) organic compound which is soluble in both water and oil. In the context of lung pathology, surfactant is a phospholipoprotein formed by Type II pneumocytes (alveolar cells). Surfactant coats the inner surface of the alveolus, and with the hydrophobic tail ends of molecules facing the air, acts to reduce surface tension, increase pulmonary compliance, and prevent the collapse of the lung (atelectasis) at the end of expiration.

Syncytium: A mass of cytoplasm surrounded by a membrane containing multiple nuclei and no internal cell boundaries. You may encounter this word used in the context of giant cell formation.

Tachypnea: Abnormally rapid breathing, a nonspecific sign in respiratory disease or infection.

Trabecula (pl. trabeculae): A supporting or anchoring strand of connective tissue. Trabeculae usually extend from a capsule into the tissue of an enclosed organ and provide structure.

Tuberculin skin test: A test for exposure to TB. It involves putting a small amount of TB purified protein derivative (which acts as an antigenic stimulus) under the top layer of skin on the inner forearm and measuring the local immune response in the form of a skin reaction (the diameter of palpable raised hardened area across the forearm perpendicular to the long axis). The most common tuberculin skin test is the Mantoux test (also known as the Tuberculin sensitivity test, Pirquet test, or PPD test). Other skin tests include the Heaf test and Tine test.

Tuberculoma/tuberculoid granuloma: A macroscopically detectable mass, sometimes mistaken for malignancy but which is instead a granuloma due to TB infection. See also *granuloma*.

Ziehl-Neelsen (ZN) stain: Named after two German doctors, bacteriologist Franz Ziehl (1859–1926) and pathologist Friedrich Neelsen (1854–1894), this is a special bacteriological stain used to identify acid-fast organisms, in particular, *Mycobacterium*. The reagents used are Ziehl-Neelsen carbol fuchsin, acid alcohol, and methylene blue. The lipid-rich cell wall of *Mycobacterium tuberculosis* makes it resistant to Gram staining but these bacilli will be bright red after ZN staining. The ZN stain will also stain other bacteria such as *Nocardia*.

Index

H

I

W

X

Z

For Product Safety Concerns and Information please contact our EU representative GPSR@taylorandfrancis.com Taylor & Francis Verlag GmbH, Kaufingerstraße 24, 80331 München, Germany

Batch number: 10777146

Printed by Printforce, the Netherlands